COMPONENTS. PRODUCTS. MACHINERY.

Vol. 8 • Issue 05 • JULY 2014 • Pages 144 • ₹ 100

ELECTRONICS

INDIA'S FIRST ELECTRONICS SOURCING MAGAZINE
www.electronicsb2b.com

BAZAAR

FROM THE PUBLISHERS OF ELECTRONICS FOR YOU

DEFENCE
ELECTRONICS
Major Business Opportunities
Come Knocking

I0468019

WHAT'S NEW IN

- Multimeters
- SMT printer systems
- Home/office inverters
- Biometric devices
- High power/lamp LEDs

IN CONVERSATION WITH
Ravinder Zutshi,
deputy managing director,
Samsung India Electronics

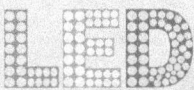  | Pgs 91-105 | ⟩⟩ **Top 20 Electronic Component Manufacturers** | ⟩⟩ **South Region Special**

Driving Growth of LED Business in India

 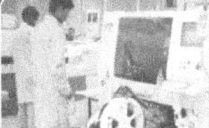   

     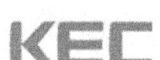

Contents

www.electronicsb2b.com

COMPONENTS. PRODUCTS. MACHINERY.

ISSN-0974-1062

ELECTRONICS BAZAAR
INDIA'S FIRST ELECTRONICS SOURCING MAGAZINE

VOL. 08 • ISSUE 04 • JULY 2014

COVER STORY

DEFENCE ELECTRONICS
Major Business Opportunities Come Knocking
28

FACTS & FIGURES
24 Some major investments in the Indian semiconductor industry

GOVT POLICIES & SCHEMES
26 Lean Manufacturing Competitiveness Scheme offers 80% subsidy to MSMEs

LEADING PLAYER
42 Top 20 electronic components manufacturers In India and leading electronics distributors in India

EMS ZONE
54 Dixon Technologies aims to be one among the top five global EMS companies

SOLAR FOCUS
64 Rooftop solar PV system: Will it be a game changer?

SMT FOCUS
68 The latest in SMT printer systems

T&M FOCUS
76 Today's multimeters are packed with next-gen features

ePOWE ZONE
82 Home/office inverters: What's new in the market

PRODUCT FOCUS
92 Latest LEDs deliver high efficacy with low thermal resistance, making products less expensive

TRENDS
102 Innovations in LED technology and wider adoption are opening up new markets

BUYERS' ZONE
104 Buy lumen, not power

105 NEW LED PRODUCTS

SECURED ZONE
86 Biometric devices are more practical and affordable

107 SOUTH REGION SPECIAL

IN THE NEWS
10 Innovative retail outlets to sell solar products
12 Japanese firms keen to participate in different electronics projects in India
14 Nehru Place crumbling under the weight of its problems: Are govt authorities to blame?
16 LEDMA revives operations by opening branch in Delhi

REGULARS
6 EDITORIAL
8 LEADING SUPPLIERS INDEX
18 NEWS
106 NEW MANUFACTURING FACILITIES
122 NEW PRODUCTS

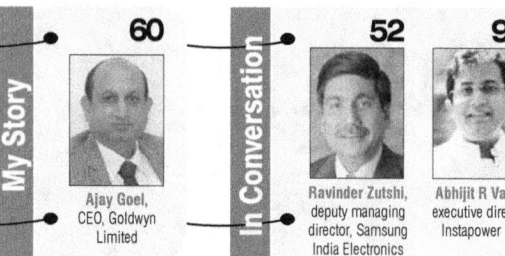

My Story
60
Ajay Goel, CEO, Goldwyn Limited

In Conversation
52 Ravinder Zutshi, deputy managing director, Samsung India Electronics
98 Abhijit R Vaish, executive director, Instapower Ltd

Editorial

COMPONENTS. PRODUCTS. MACHINERY.   www.electronicsb2b.com

ELECTRONICS
INDIA'S FIRST ELECTRONICS SOURCING MAGAZINE   BAZAAR

EDITOR	: Rahul Chopra
EDITORIAL CORRESPONDENCE	: Editorial Secretary Phone: 011-26810601 E-mail: editsec@efyindia.com (Technical queries: efylab@efyindia.com)
SUBSCRIPTIONS & MISSING ISSUES	: Phone: 011-26810601 or 02 or 03 E-mail: support@efyindia.com
BACK ISSUES, BOOKS, CDs, PCBs etc.	: Kits'n'Spares, New Delhi Phone: 011-26371661, 26371662 E-mail: info@kitsnspares.com
EXCLUSIVE NEWSSTAND DISTRIBUTOR	: IBH Books & Magazine Distributors Ltd, Mumbai Phone: 022-40497401, 40497402, 40497474, 40497413; Fax: 40497434 E-mail: circulations@ibhworld.com
ADVERTISEMENTS NEW DELHI (HEAD OFFICE)	: Ph: 011-26810601 or 02 or 03 E-mail: efyenq@efyindia.com
MUMBAI	: Ph: 022-24950047, 24928520 E-mail: efymum@efyindia.com
BENGALURU	: Ph: 080-25260394, 25260023 E-mail: efyblr@efyindia.com
CHENNAI	: Ph: 09916390422 E-mail: efyenq@efyindia.com
HYDERABAD	: Ph: 09916390422 E-mail: efyenq@efyindia.com
KOLKATA	: Ph: 08800094201 E-mail: efyenq@efyindia.com
PUNE	: Ph: 09223232006 E-mail: efypune@efyindia.com
GUJARAT	: Ph: 09821267855 E-mail: efyahd@efyindia.com
CHINA	: Power Pioneer Group Inc. Ph: (86 755) 83729797, (86) 13923802595 E-mail: powerpioneer@efyindia.com
JAPAN	: Tandem Inc., Ph: 81-3-3541-4166 E-mail: tandem@efyindia.com
SINGAPORE	: Publicitas Singapore Pte Ltd Ph: +65-6836 2272 E-mail: publicitas@efyindia.com
TAIWAN	: J.K. Media, Ph: 886-2-87726780 ext. 10 E-mail: jkmedia@efyindia.com
UNITED STATES	: E & Tech Media Ph: +1 860 536 6677 E-mail: veroniquelamarque@gmail.com

The hopes of the entire electronics industry are riding on the new government at the Centre. Going by reports in the Press, it seems that the new government is taking the challenges faced by the electronics industry seriously, and realises the importance of boosting indigenous manufacturing and exports to revive the sector.

The questions that one could ask here are: What track will the Ministry of Communications and IT follow? What should be the agenda of the new government to boost electronics manufacturing and attract investments? Will it be able to modify the tax issues that act as a deterrent to the growth of the industry? While these seem to be the major issues that the new government needs to address on priority, there are a host of others, too, that require immediate attention.

The optimistic electronics industry awaits faster implementation of the National Policy on Electronics, as growth in this sector, especially in manufacturing, has not taken off yet.

Industry giants recommend that besides the key issues to be addressed, the new government needs to focus on creating a national electronics mission with adequate manpower to implement the policy. The Electronics Development Fund needs to be set up immediately to encourage design and manufacturing value addition in India and to encourage entrepreneurship within the industry. The Preferential Market Access (PMA) policy, a bold and effective step to ensure value-added manufacturing in India, should be revived immediately. This policy can become a strong catalyst to boost domestic manufacturing of electronics, while attracting investments.

Another important issue the new government should act upon is to involve all the state governments in this entire exercise, and encourage them to facilitate investments and fast clearances.

Inaction or the slow pace of approvals while releasing funds for projects under the Modified Special Incentive Package Scheme and electronics manufacturing clusters is leading to delays in setting up the proposed infrastructure. Here again, faster clearances are required.

The new government should also immediately address the concerns of the big investors in electronics manufacturing in India. For example, Nokia's largest cell phone factory worldwide, located in Tamil Nadu, is closing down. This has led to the Flextronics factory, which supplies its products to Nokia, to also shut down. These two electronics heavyweights had not only created an electronics ecosystem in the state, but had also shown confidence in the Indian policies. The closing down of these factories will have a major negative effect on the industry, and will lower international investor confidence in India.

So the list of the electronics industry's woes is long. It will be interesting to see how the new government addresses these challenges to take the industry forward.

Srabani Sen

Srabani Sen
Executive Editor, Electronics Bazaar
srabani.sen@efyindia.com

SUBSCRIPTION RATES FOR ELECTRONICS BAZAAR			
Period Year	News-stand Price (₹)	You Pay (₹)	Overseas
Five	3000	2080	—
Three	1800	1340	—
One	600	475	US$ 65

Kindly add ₹ 50/- for outside Delhi cheques.
Please send payments only in favour of **EFY Enterprises Pvt Ltd,** payable at Delhi
Non-receipt of copies may be reported to support@efyindia.com, mentioning your subscription number

 **HONGFA RELAY**

 **MILLENNIUM SEMICONDUCTORS**

THE LEADING RELAY R&D AND MANUFACTURING BASE GLOBALLY

Xiamen Hongfa Electroacoustic Co. Ltd is the first largest relay manufacturer in China and one of the leading relay manufacturers and suppliers in the world. It produces a wide range of products, including relays, low-voltage devices, precision components and automatic equipments. In particular, it has over 160 types of relay categories, including power relays, latching relays, automotive relays, signal relays, industrial & electrical appliance relays, safety & electrical appliance relays, solar relays and hermeticallysealed relays, which provide over 40,000 regularly-used relay specifications.

POWER RELAY

UPS, Inverter, Control panel, Voltage stabilizer

HF3FA (215)
- 1 pole: 15A 125VAC, 10A 250VAC/28VDC
- Product in accordance to IEC 60335-1 available

HF3FF (215)
- 1 pole: 10A 277VAC/28VDC

HF14FW(136)
- TV-8, 20A switching capability (1 pole configuration)
- 4kV dielectric strength
- Meeting VDE 0700, 0631 reinforce insulation

UPS, Inverter · Railways and Air-con. · Timers/Relay module

HF105F-1(136)
- 40A switching capability
- 4kV dielectric strength (coil to contacts)
- Unenclosed, Plastic sealed and dust protected types available

HF102F
- Heavy load up to 5000VA
- Ideal for motor load (80A High inrush current)
- PCB & QC layouts available

HF33F
- 1 pole: 10A 125VAC, 5A 250VAC/30VDC
- Creepage distance: 8mm (NO to NC)
- High sensitivity 200mW available

Lighting application · UPS · PLC Controller · Air-con.

HF46F
- I pole. 5A 277VAC/30VDC at 85℃
- Surge withstand voltage up to 10kV
- High sensitivity: 200mW

HF115FD
- 16A switching capability
- Creepage distance: 10mm
- Updated version of HF115F 1 pole equipped with automatic production line

HF49FD
- 3kV dielectric strength between coil and contacts
- Slim size: width 5mm, height 12.5mm
- High sensitive: min. 120mW

HF118F(136)
- 1 pole: 10A、2组 (2 pole) 5A
- dielectric strength : 5kV
- Creepage distance >8mm
- Reflow soldering available

Telecom, EPABX, Mobile tower

HFD27
- High contact capacity 2A 30VDC
- Matching 16 pin IC socket
- Epoxy sealed for automatic-wave soldering and cleaning

HFD23
- 1 Form C configuration
- 2A switching capability
- High sensitive: 150mW
- Plastic sealed type available

HFD3
- Surge withstand voltage up to 2500VAC, meets FCC Part 68 and Telecordia
- SMT and DIP types available
- Single side stable and latching type available

Control panel

HF18FF / HF18FH
- 7A switching capability (2C, 3C type)
- Various terminals, test button available
- Gold plated contact available
- 2 to 4 pole configurations

HF13F
- 15A switching capability (1 Form C)
- Conform to the CE low voltage directive
- 1 & 2 pole configurations
- Various terminals available

LEADING SUPPLIERS' INDEX

Products	Page No.
Batteries & Power Supplies	
Balaji Powertronics	131
Indtronies	141
Intex Power Electronics	140
Minmax Technology Co. Ltd	31
Montu Electronics	140
Mornsun Guangzhou Science & echnology Co.Ltd.	51
Nippon India (D)	135
Prime Products	135
Rishabh Instruments Pvt Ltd	23
Sakthi Accumulators Pvt Ltd	136
Servokon Systems Pvt Ltd	126
Southern Batteries P Ltd	57
Sunlight	141
Temco Electricals & Electronic Industries	136
Uniq Power Solutions	134
Upsinverter.com	125

Products	Page No.
Cabinets, Enclosures & Accessories	
Radisson Instruments	140
Shrey Plastic Moulders	141

Products	Page No.
Components (Including Active & Passive)	
Ceepee Electronics	134
Cirkit Electro Components Pvt Ltd	87
Componix India	33
CRT Comptex Pvt Ltd	130
Electronika Sales Corpn.	75
International Rectifier Hong Kong Ltd	19
LWI Electronics Inc.	21

Products	Page No.
Maruti Electronics	140
Millenium semiconductors	7
Mouser Electronics (Hong Kong) Ltd.	144
Murata Manufacturing Co. Ltd.	27
Parshwa Components	140
Precision Electronic Components MFg CO.	45
Progressive Technologies	135
Sancon India Pvt. Ltd.	137
Schurter Electronics (I) Pvt Ltd	127
Silicon Power Electronics	132
Swingtel Communications Pvt. Ltd.	2

Products	Page No.
Industrial & Manufacturing Equipment	
Accurex Solutions (P) Ltd.	128
Accurex Solutions (P) Ltd.	129
Amptronics Systems Pvt Ltd.	126
ASYS Group Asia Pte Ltd.	71
Drive Technologies	59
Juki India Pvt Ltd	15
Nordson India Pvt Ltd	13
Omron Automation Pvt. Ltd.	142
Sumitron Exports Pvt Ltd	37
Transtechnology India Pvt Ltd	69

Products	Page No.
Materials (Including Chemicals & Consumables)	
HK Wentworth (India) Pvt Ltd	9
Premier Industries	134
Static Systems Pvt Ltd	133, 138

Products	Page No.
Optics & Optoelectronics	
Avni Energy Solutions Pvt Ltd	101

Products	Page No.
Boisi	97
Brisk Electronics	99
Everlight Electronics India Pvt Ltd	93
Goldwyn Ltd.	95
Kwality Photonics P Ltd	94, 96
Melux Control Geaars Pvt Ltd.	97
Nura Enterprises	99
Pyrotech Electronics Pvt Ltd.	99
Recom Asia Pvt Ltd	59
Temco Electricals & Electronic Industries	136

Products	Page No.
PCBs, Assemblies & Sub Assemblies	
Buljin Elemec Pvt Ltd	126
CIPSA TEC India Private Ltd	130
Digital Circuits Pvt Ltd.	1
SMD Electronics Pvt.Ltd.	141
Sulakshana Circuits Ltd.	132

Products	Page No.
Plugs, Sockets & Connectors	
Max Electronics	141

Products	Page No.
Reseller and Distributors	
LWI Electronics Inc.	21
Millenium semiconductors	7
Mouser Electronics (Hong Kong) Ltd.	144
Nichani Electronics	130
Swingtel Communications Pvt. Ltd.	2

Products	Page No.
Safety & Security Products	
Bosch Ltd.	11
Global Tele Communications	140

Products	Page No.
Solar Products	
Borg Energy India Pvt Ltd.	5
Hitech Solar Appliances	140
Vidyut Power Systems	136

Products	Page No.
Telecom Products	
Global Tele Communications	140

Products	Page No.
Test & Measurement Equipment (Including Indicators & Monitors)	
Agilent Technologies India Pvt. Ltd.	25
Pacific Electronics (P) Ltd.	132
Rohde & Schwarz	35

Products	Page No.
Trade Shows and Events	
Elcina SES	83
Gujarat Manufacturing Show 2014	63
IPCA 2014	89
Light India	51

Products	Page No.
Transformers	
Jai Mata Electronics	140

Products	Page No.
Wires & Cables	
Silvertone Trading Co.	140

Products	Page No.
Miscellaneous	
Savi Plastic & Electronics Pvt Ltd.	141
Persang Entertainment Pvt Ltd.	139

Page numbers subject to final dummy corrections

ORGANISATIONS COVERED IN THIS ISSUE

Organisation	Page No.
3B Semiconductors Pvt Ltd	48
4G Identity Solutions	90
Accurex Solutions Pvt Ltd	108
Agilent Technologies	30,80,122
Alcon Electronics Pvt Ltd	42,46
Alfa Electronic Components	48
All Delhi Computer Trader Association (ADCTA)	14
Amptronics Systems Pvt Ltd.	115
Anu Solar Pvt Ltd.	10
ASYS Group GMBH	74
Avni Energy Solutions Pvt Ltd	110
Axis Communications	124
Base Batteries	123
BG LI-IN Electricals Ltd	42,44
Bharat Dynamics Limited (BDL)	29
Bharat Electronics Limited (BEL)	29
Bharat Electronics	23
BORG Energy India Pvt Ltd.	109
Bosch Security Systems	111
Bridge To India	64
Buljin Elmec Pvt Ltd.	119
CEE PEE Electronics	48
Cermet Resistronics Pvt Ltd	42,47
Cirkit Electro Components Pvt Ltd.	48
Componix India.	50
Convergence Power Systems Pvt Ltd	84
Cosmo Ferrites Ltd	42,44
DEK	68
Deki Electronics Ltd.	42,43
Department of Information Technology (DeitY)	12,30,53
Desai Electronics Pvt Ltd.	42,44
Digital Circuits Pvt Ltd	116
Dixon Technologies India Pvt Ltd	54,55,56,58
ELCINA.	29
Elektronika Sales Pvt Ltd.	114
EMS Technologies Pvt Ltd.	70
Epcos India Pvt Ltd.	42,43

Organisation	Page No.
eSSL Pvt Ltd.	86,88
Essmart Global	10
Eveready Industries	20
Everlight Electronics Co Ltd	113
F-Secure.	20
FCI.	123
Flir Systems Inc.	80
Fluke	76,78
Fortuna Impex Pte Ltd	86,88
Fortune Marketing Pvt Ltd	88
Foxconn India	22
Gautam Solar Pvt Ltd	106
GE (General Electric) Power Electronics.	23
General Industrial Controls Pvt Ltd.	42,44
Genius Electrical & Electronics Pvt Ltd	42,44
Genus Power Infrastructures Ltd.	82,84
GlacialLight.	105
Globe Capacitors Ltd.	42,43
Goldwyn Limited	60,61,62
Good Will Instrument Co Ltd	78
Greenstar.	104
Gujarat Poly AVX Electronics Ltd	42,47
GW Instek	78
Harman International	106
Hexi Electronic Equipment Co Ltd	74
Hindustan Aeronautics (HAL)	29
Hioki EE Corporation	80
HK Wentworth (India) Pvt Ltd (Electrolube)	112
Honeywell Automation India Ltd	106
Honeywell Process Solutions (HPS)	20
Ideal Industries India Pvt Ltd	78,79
Incap Ltd.	42,44
Indium Corporation	124
Infosys	104
Instapower Ltd.	98
Intellix Security Solutions.	88
Juki Automations Systems Corporation	72
Keltron Component Complex Ltd	42,43
Kwality Photonics Pvt Ltd	94,96,120

Organisation	Page No.
LED Products Manufacturers Association (LEDMA)	16
Luminous Power Technologies	85,124
LWI Electronics Inc.	115
Matrix Comsec India Pvt Ltd.	88
MEL Systems and Services Ltd.	29,30
Metro Electronic Products Ltd	78,79
Micro, Small and Medium Enterprises (MSME)	18
Micromax	20
Microtek International Pvt Ltd	84
Millennium Semiconductors.	50
Ministry for New and Renewable Energy (MNRE)	10,18,22,65,66,67
Ministry of Defence	29
Mornsun.	123
Mouser Electronics	117
Nichani Electronics	113
Nichia Chemical India Pvt Ltd.	92,94
Nordson India Pvt Ltd	107
NSIC.	29,38
NTPC Ltd.	106
Nura LED.	110
NXP Semiconductors	105
Omniscient Associates	120
Osram Opto Semiconductors	105
Pace Power Systems	23
Pacific Electronics Pvt Ltd	114
Philips Lumileds.	92,94,102,104,105
Philips.	104
Prismatic Engineering Pvt Ltd	42,46
Progressive Technologies.	112
Rabyte Electronics Pvt Ltd	50
Radius Industries	47
Rangsons Electronics Pvt Ltd.	30
Rank Infotech	47
RECOM Asia Pte Ltd.	116
Renesas Electronics Singapore	23
Rishabh Instruments Pvt Ltd	78
Rohde & Schwarz.	122

Organisation	Page No.
ROHM.	123
Sakthi Accumulators Pvt Ltd (formerly Sakthi Electronics).	119
Samsung India Electronics	52
Samsung	68
Saurya EnerTech	10
Schwing Stetter India.	106
Secureye	88
Solid State Systems Pvt Ltd	42,46
Southern Batteries Pvt Ltd.	117
Speedline Technolgies Inc.	72
Speedofer Components Pvt Ltd	42,46
Static Systems Pvt Ltd.	118
STMicroelectronics	123
Su-Kam Power Systems Ltd.	10,64,82,84
Sulakshana Circuits Ltd.	109
Swelect Energy Systems Ltd.	106
Swingtel Communications Pvt Ltd.	50
Taiwan Semiconductor Lighting Company (TSLC).	92,94,96
Tandon Group.	29
Tata Power Solar	64
Temco Electricals & Electronic Industries.	118
Texas Instruments	122
Thakor Electronics Ltd.	42,46
Toshiba Lighting.	104
Uniq power solutions	111
upsINVERTER.com.	82,84,85
Victor Component Systems Pvt Ltd	42,46
Vishay Components India Pvt Ltd	42,43
Vishay Intertechnology	122
Watts Electronics Pvt Ltd.	42,46
Welspun Renewable Energy	106
Welspun Renewable Energy	23
World Trade Organisation (WTO)	18
Yamaha Motor Co Ltd.	124
Yamaha.	70

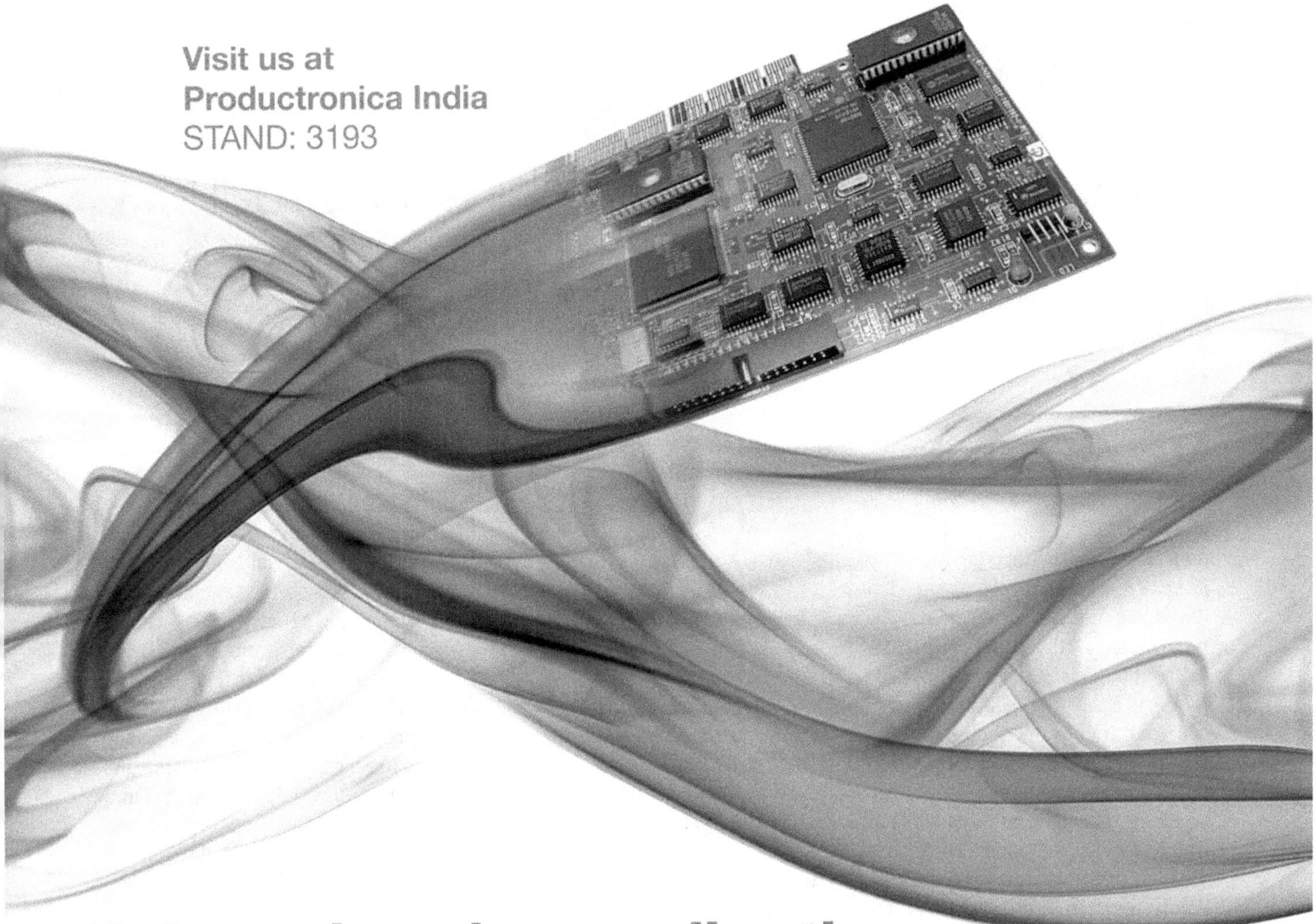

Innovative retail outlets to sell solar products

While MNRE's retail initiative has not been very fruitful, with many Akshay Urja Shops rendered non-functional, several solar manufacturers and traders have come up with innovative ways to sell solar products

By EB Bureau

Tarun Kapoor, joint secretary, Ministry of New and Renewable Energy (MNRE), recently said at a seminar that India's lack of retail outlets for solar products is hampering its sales to a large extent. "We need to develop a mechanism wherein solar products are available in the open market so that consumers can buy these products easily," said Tarun Kapoor.

According to the manufacturers, though solar products such as lanterns, emergency lights, PV modules, water heaters, etc, are available at the retail shops both in cities and villages, their numbers are quite less. For example, in Delhi, such products are available at shops in electrical markets like Chandni Chowk.

Retail shops proliferate in rural areas

Till now, the government was solely responsible for boosting the use of solar products. But today, observing the demand for these products, particularly in rural areas and regions where power outage is rampant, manufacturers are either opening their own retail shops or are tying up with retail outlets. These products are usually available in retail shops that deal with electrical products.

"Since solar products do not get categorised as part of consumer electronics, we cannot expect them to be available at the typical consumer electronics shops. Although there are retail shops in cities and villages, their number is quite meagre. But with the demand rising, the number of such shops will grow," says Dr Satyendra Kumar, chairman, Saurya EnerTech. "Manufacturers are deliver-

ing their products to the market through their distributors, who are taking them to the retail shops. Retailers also approach the manufacturers directly for such products," he adds.

Government initiative

In order to make solar energy products easily available and to provide easy after sales repair services, in 1995, MNRE established the Aditya Solar Shops, which were later renamed as Akshay Urja Shops, in major cities. These shops were established by the state nodal agencies (SNAs), manufacturers' associations and some NGOs. Later, private entrepreneurs were also allowed to establish these shops.

According to Dr Satyendra Kumar, the Akshay Urja Shops failed to perform well. Many of them are either not doing good business or have closed down. For example, the Akshay Urja Shop that was set up near Madras Hotel in New Delhi has now closed down. The government is seeking applications to open this shop again. In many other cities and towns like Warangal, Jaipur, Bhopal, Dehradun, Guwahati, Patna and Chennai, and some districts in Uttar Pradesh and East Sikkim, these shops have been closed.

"Manufacturers find it difficult to be a part of this scheme as the pa-

rameters that need to be fulfilled are so tough that quite a number of them do not qualify," says Dr Satyendra Kumar.

Entrepreneurial initiative

Essmart Global has converted many local kirana shops into retail shops for solar products. Essmart buys products like solar lanterns and water filters in bulk from manufacturers, runs a sample test on them, and then places them at kirana shops for local customers to access these products. The owner of the kirana shop gets 8-10 per cent from the profit margin.

Bengaluru-based Anu Solar Pvt Ltd, a leading manufacturer and trader of solar products, has opened several solar retail shops. Called Anu World, these sell various products ranging from solar water heaters and solar powered calculators, to caps with fans that run on solar power and solar power-based storage batteries and inverters. These stores sell the products at MNRE subsidised rates, as the company is a channel partner of MNRE.

In an innovative move, Su-Kam Power Systems has tied up with Dhampur Sugar Mills to sell solar products in rural pockets across Uttar Pradesh (UP) and Uttarakhand. Su-Kam's products will be sold at 'e-haats' run by Dhampur Sugar, which is a leading integrated sugar company in India. E-haats are retail outlets present in over 350 villages in UP and Uttarakhand, and are one-stop shops for agricultural equipment, bio-fertilisers, soil testing and other agri services.

"Through this initiative, Su-Kam will sell technologically advanced solar products that offer people value for money, thereby helping them in both generating and saving energy through solar power usage," says Dhananjay Sharma, general manager, solar networks, Su-Kam. EB

DCN multimedia Conference System

Inform. Impress. Inspire.

Introducing the brand new DCN multimedia Conference System from Bosch, an exciting new development that will inform, impress and inspire. With high-resolution capacitive touchscreens, an unobtrusive microphone and extremely high sound quality, meeting participants become truly engaged. It's the first ever IP-based conference system built on the OMNEO media networking architecture and operates on standard Ethernet networks. This makes the system extremely scalable and flexible, and at the same time ensures cost-effective installation and maintenance.
http://in.boschsecurity.com

BOSCH
Invented for life

For more information contact:
Bosch Limited, Security Systems Division,
P B No. 3000, Adugodi, Bangalore - 560030
Ph: +91(80)67528378 Fax: +91(80)67528263.
Email - boschsecuritysystems@in.bosch.com

Japanese firms keen to participate in different electronics projects in India

Japan has sought strong intervention by the Indian government to prevent certain chronic bottlenecks, subject to which Japanese firms would be keen to invest in India's electronics manufacturing ecosystem

By EB Bureau

Japanese organisations are considering India as a strategic hub and are keen to participate in different projects across the country, said Tamaki Tsukada, minister, economic development, embassy of Japan, at a round table discussion on 'Electronics Manufacturing Industry—Challenges and the Way Forward', organised by the PHD Chamber of Commerce and Industry (PHDCCI), along with the Department of Information Technology (DeitY), on May 26, 2014, in New Delhi.

Tsukada shared the Japanese government's perspective on three issues pertaining to the electronics manufacturing ecosystem in India—the complicated tax system in India, lack of infrastructure, and delays in procedures and processes.

"Japan proposes five items for consideration by the Government of India, all of which need strong intervention by the government—support for all the sectors of the electronics manufacturing industry, support measures for subsidising the ventures, availability of skilled HR at all levels of the management, simplification of the tax system, and tariff cuts in order to be integrated into the global supply chain," he said.

R&D: The need of the hour

Dr Ajay Kumar, joint secretary, DeitY, who was the chief guest at the event, said that R&D is of utmost importance in the fields of electronics manufacturing; hence, he emphasised the need for skills development programmes to be undertaken by the ministry to create skilled hands that meet industry needs. "DeitY is

working closely to develop better relations with Japan to attract investments from Japanese organisations. The Ministry of Communication and Information Technology has developed investor friendly policies to attract investment and to help in the growth of the electronics industry," he said. He also said that the government is focusing on building up globally competitive electronics design skills to meet both the country's needs and serve the international markets.

Haryana: An investor friendly state

Salil Narang, general manager, investment promotion centre, HSIIDC, talked about Haryana being an investor friendly state with state-of-the-art infrastructure and ample technical manpower. He said that Haryana has the highest per capita income and is the largest producer of automobiles in the country. "To enable electronics manufacturing in India, the state has developed entirely different SEZs. The Haryana government policies have attracted different IT and electronics organisations to invest in a politically stable environment with a GDP of about Rs 64.6 billion," he said.

Focus more on design and manufacturing

Sanjeev Gupta, chairman, ICT committee of PHDCCI, said that India needs to concentrate and focus more on designing and manufacturing global products

(L-R) Tamaki Tsukada, minister, economic development, embassy of Japan, Sanjeev Gupta, chairman, ICT committee of PHDCCI, Dr Ajay Kumar, joint secretary, DeitY and Salil Narang, general manager, investment promotion centre, HSIIDC, at a round table discussion on 'Electronics Manufacturing Industry— Challenges and the Way Forward,' in New Delhi

and exploring export markets. "Not only should India leverage its strengths in software to build products of high complexity, but also work towards manufacturing medium-volume products for the global market. Further, the Indian industry should focus on inventing mass-products that matter to the rural and 'bottom of the pyramid' segments," he said.

Sanjeev Gupta further said that the demand for appliances and energy efficient consumer electronics is huge and can be explored by the Indian electronics industry. Increasing R&D intensity should be the joint effort of both the government and industry. With such strategies, the Indian electronics industry would excel both in domestic and international markets.

The round table focused on addressing the issues and challenges that prevail in the process of implementing the policies of DeitY. It identified the effective courses of action to be taken to ensure that these policies result in the growth of the electronics manufacturing industry. EB

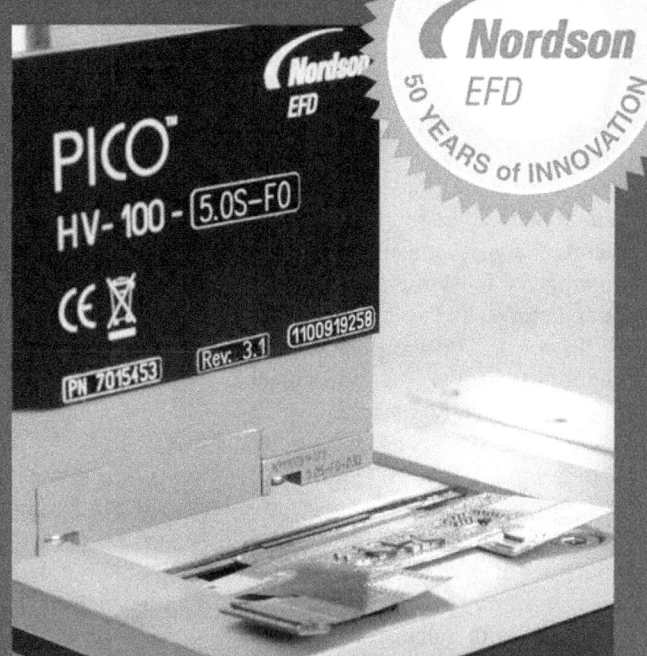

Nehru Place crumbling under the weight of its problems: Are govt authorities to blame?

Despite massive volumes of business being churned out of this IT hub, the place lacks basic amenities

By Richa Chakravarty

Recently, the US trade representative (USTR) listed Nehru Place as among the 30 most notorious IT markets in the world. Apart from being a haven for software piracy and a flourishing grey market, Nehru Place is one of the biggest commercial hubs of India and is the largest market for IT products in Asia.

Established in 1969, this business centre was named Kalkaji Complex and subsequently renamed Nehru Place in the 1980s, after India's first prime minister (PM) Jawaharlal Nehru. Out of the 15 district centres in Delhi, Nehru Place is the biggest, with more than 1200 channel partners operating from here and facilitating business worth millions, every day.

With 102 buildings, Nehru Place houses nearly 20,500 offices. Despite such massive volumes of business being churned out of this IT hub, the place lacks basic amenities like adequate parking, sanitation, maintenance of emergency exit systems, etc.

Electronics Bazaar took a closer look at the problems faced by traders in Nehru Place. We surveyed the market and spoke to the All Delhi Computer Traders Association (ADCTA) president, Mahinder Aggarwal, on the issues that concern traders.

Limited parking area

Nehru Place currently has seven authorised parking areas along with the provision for underground parking, with the capacity to accommodate 471 cars, 672 cycles and 600 scooters. But this parking area was built in 1975. "Despite the fact that the number of offices and shops in

A regular sight at Nehru Place

Mahinder Aggarwal,
president, ADCTA

Nehru Place has increased tremendously, the parking area has never been expanded to accommodate the increasing number of vehicles. Poor maintenance and lighting conditions in the parking area are another issue that concern the traders," says Mahinder Aggarwal.

The Delhi Metro Rail Corporation (DMRC) has its own parking area; however this is not sufficient for the needs of the traders. Most of the parking done here is illegal. In such a scenario, traders have to park their cars in illegal passages. Visitors and customers are the most affected as they do not get parking space in Nehru Place. "We have to park our vehicles away from the main market," shares a visitor. "But with the connectivity of the Metro line to Nehru Place, travelling to this place has become easier, though parking is still a big problem."

Illegal hawkers: The golden goose?

The most interesting aspect to Nehru Place is the long row of pavement vendors right in front of the showrooms. From pirated CDs of software and movies, to electronic gadgets, one can find everything on the pavement.

This is one of the biggest problems for the traders and channel partners occupying the showrooms and offices, who find it tough to compete with the throw-away prices that the hawkers offer for a pirated product. As per the court's orders, only 114 hawkers can sit in Nehru Place (spread across various locations—outside the metro station, on the main pavement or at the bus stop). However, more than 700-800 hawkers can be found on the main pavement itself.

"Each hawker pays Rs 1000 per day to the police officer in-charge to conduct business on the pavement. For the government authorities, this is one of their biggest sources of revenue and who would like to kill the goose that lays the golden eggs? On a single day, it is estimated that nearly Rs 0.1 million goes into the pockets of the authorities for illegally allowing pavement vendors to carry on their business from Nehru Place," informs Mahinder

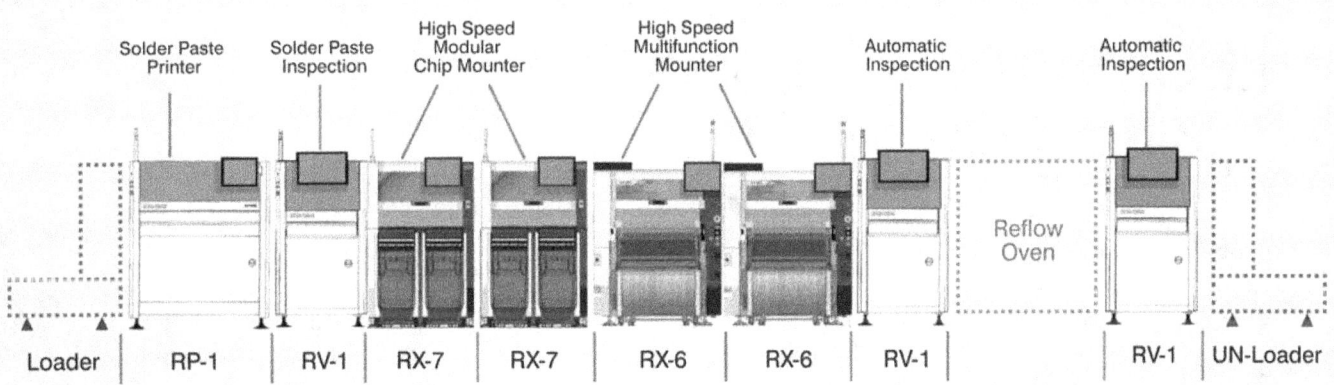

Aggarwal. He further adds, "We, as an association, have made several complaints but no action has been taken; rather, the traders and associations' members receive threatening calls and warnings to mind their own business."

Inadequate security

Despite housing several buildings, the market complex has no proper security system. A sufficient number of CCTV cameras has not been installed and most of the traders complain that these cameras do not operate most of the time. The ongoing third phase of the DRMC has also created chaos at the market. Traders complain that due to the ongoing metro work, the cables of the security cameras often get cut. Also, the police officials as-signed to the market are just not sufficient to handle any untoward activity. "There are hardly 8-10 police officers present in this market. Keeping in mind the huge size and requirements of this complex, we need more than 100-120 police personnel to ensure adequate security," says a trader.

Sanitation and hygiene

This is another sore point for the traders at Nehru Place. Not only are there piles of garbage lying around in the complex but no maintenance has been done on the buildings for years. Most of the corridors do not have proper exits and are not well lit. Surprisingly, a big business centre like Nehru Place does not have the basic amenities of enough washrooms and water coolers. Though the big companies and offices have their own facilities, the traders do not have any such privileges. "The Delhi Development Authority (DDA) undertook the responsibility of cleaning and beautification of Nehru Place for two years, however, it has been 14 years since it began and in the name of cleaning, we have piles of garbage lying around. There are too few washrooms in this business complex to serve the number of people that Nehru Place holds. Sanitation and hygiene are other reasons why people prefer not to visit this market. Several times we have approached DDA to provide us land or space so that we can build washrooms or install water coolers, however our request seems to be falling on deaf ears," complains Mahinder Aggarwal. ☒

LEDMA revives operations by opening branch in Delhi

With the inauguration of its new branch office, the association has registered a 10 per cent growth in its membership, and will soon elect a new president

By Richa Chakravarty

LED products Manufacturers Association (LEDMA), the only association catering to the needs of the LED industry in India, has given its operations a boost by inaugurating a branch in Delhi in April 2014.

Established with much fanfare in 2010, LEDMA was formed by Dr Ramana Rao, chairman and managing director of MIC Electronics with the sole objective to promote awareness about solid state lighting and introduce energy efficient lighting systems in India. The association aimed to work towards accelerating manufacturing in the LED sector and to address the issues hampering its growth. However, LEDMA soon became inactive and did not succeed in generating much awareness over the last three years.

Revival of LEDMA

Witnessing the current growth of the LED industry in India, LEDMA decided to revive the association and has opened a branch in Delhi. It will soon select its new president. "Delhi being a central place, the association will operate from here. We have not yet decided upon the location of LEDMA's headquarters, but we may make the Delhi branch our headquarters," says Surinder Kumar Singla, secretary, LEDMA.

"Our foremost priority is to infuse some life into the association. We have formed a core committee of eight members, as well as sub-committees that are already working and will eventually decide the president for the association," adds Surinder Kumar Singla.

LEDMA is planning to create regional branches that will focus on the north, east, west and south. With the inauguration of LEDMA's new branch, the association has registered a 10 per cent rise in its membership and now has 45 members. The association will be fully operational in three months, with the revival of its website.

LEDMA plans to handhold SMEs while they continue to contribute towards indigenous manufacturing. "Our aim is to help the SMEs in all possible ways, whether through policies or by providing them the right infrastructure like testing labs, creating quality audits for them, etc," says Surinder Kumar Singla. ☒

    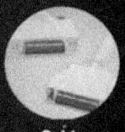 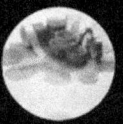  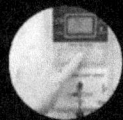 

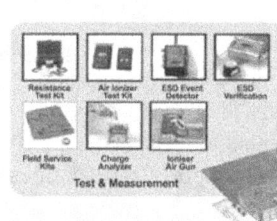

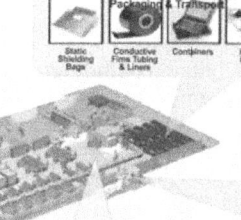

INDUSTRY NEWS ▶▶

Commerce ministry imposes anti-dumping duty on international solar panels

The Directorate General of Anti-Dumping (DGAD) has decided to impose a 'dumping' duty on the solar panels imported from the US, China, Malaysia and Taiwan, much to the ire of the Ministry of New and Renewable Energy (MNRE) and several state governments. However, the decision was hailed by the Indian solar panel manufacturers who had earlier filed a petition urging the Indian government to impose anti-dumping duties on imports from the above-mentioned countries. India has reportedly found solar products sourced from the US, China and Taiwan, dumped in the local market, which has caused 'material injury' to domestic manufacturers.

The DGAD, in its final findings, has recommended anti-dumping duties of up to $0.48 per watt on solar cells coming from the US, $0.81 per watt on those from China, and up to $0.62 per watt and $0.59 per watt from solar panels from Malaysia and Taiwan, respectively. This move may even motivate foreign producers to come and set up operations here and create local jobs. As per the reports, the commerce ministry is considering imposing a dumping margin range of 50-60 per cent on the US and 100-110 per cent on China, which is the largest exporter of solar cells worldwide. The ministry has identified 58 manufacturers, mostly from China, followed by Taiwan, Malaysia and the US, involved in the case of dumping filed by a group of domestic manufacturers two years ago.

New IT minister says first priority is to restore investor confidence

Ravi Shankar Prasad, who was recently appointed as the Minister for Communications and IT, said his priority would be to improve investor sentiments and make the functioning of the sector transparent. He told the media that there would be special emphasis on improving the overall quality of telecom services and promoting broadband roll out, especially in rural areas. The telecom sector has not attracted any major investments in the past few years. With the incumbent operators' profits declining, partly due to high regulatory costs, some of the newer players were forced to exit the sector.

WTO to examine India-US solar trade dispute, yet again

The World Trade Organisation (WTO) dispute settlement board will re-examine the spat between the two countries a second time after India decided to maintain its earlier stand that only locally made products be used in its solar industry. Hinting that consultations with India over opening up its solar market further had reached no logical conclusion, the United States has now demanded the WTO step in and create a panel to review the dispute. Meanwhile, India has expressed disappointment over the move, saying that "...the United States has chosen to litigate rather than negotiate." India further said that it had taken part in consultations with the US with an open frame of mind and that it had clearly expressed its willingness to explore all options to settle the dispute amicably.

Kalraj Mishra takes charge of MSME ministry

Kalraj Mishra has taken over as the Union minister for Micro, Small and Medium Enterprises (MSME) in the new government at the Centre. While interacting with the media, the minister highlighted the contribution of the ministry in creating jobs for the youth and opportunities for educated unemployed people in backward regions.

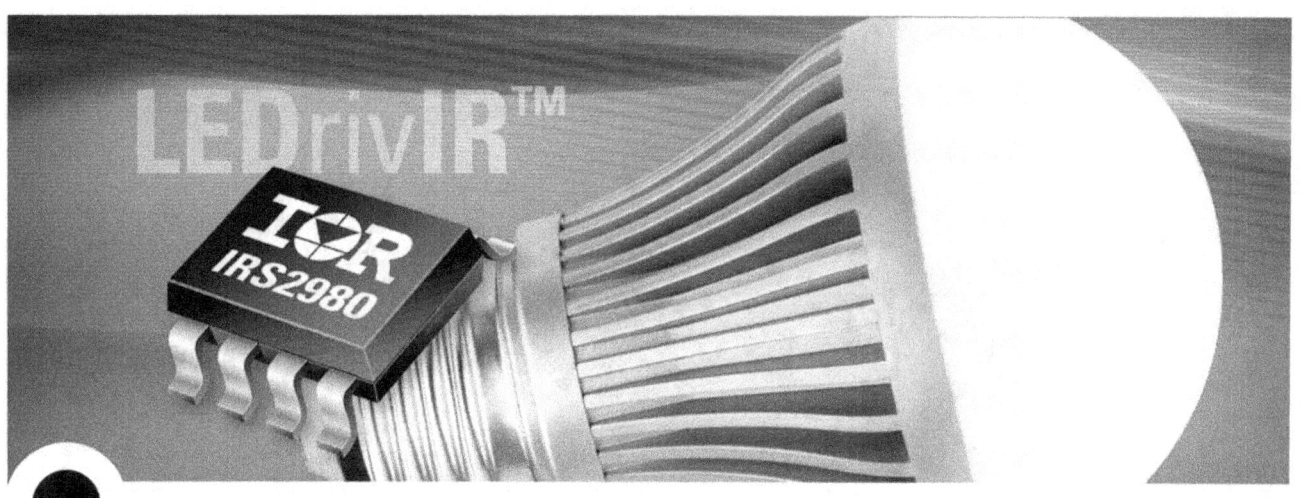

LEDriv**IR**™

IOR
IRS2980

High-Voltage Buck Control ICs for Constant LED Current Regulation

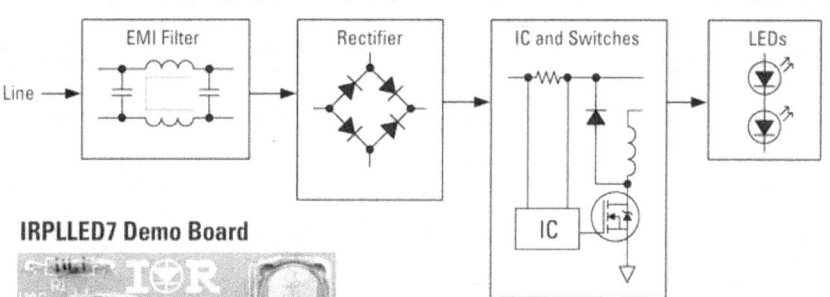

IRPLLED7 Demo Board

LEDriv**IR**™
IRS2980

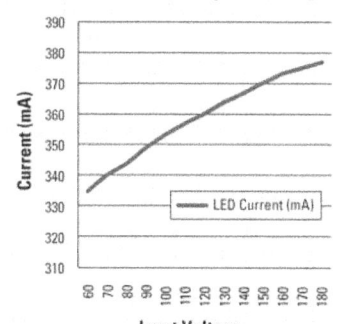

**IRPLLED7 Demo Board
LED Current vs Input Voltage**

IRS2980 Features

- Internal high voltage regulator
- Hysteretic current control
- High side current sensing
- PWM dimming with analog or PWM control input
- Free running frequency with maximum limiting (150kHz)

IRS2980 Benefits

- Low component count
- Off-line operation
- Very simple design
- Inherent stability
- Inherent short circuit protection

Demo Board Specifications

- Input Voltage 70V to 250V (AC)
- Output Voltage 0V to 50V (DC)
- Regulated Output Current: 350mA
- Power Factor > 0.9
- Low component count
- Dimmable 0 to 100%
- Non-isolated Buck regulator

Part Number	Package	Voltage	Gate Drive Current	Startup Current	Frequency
IRS2980S	SO-8	450V	+180 / -260 mA	<250 µA	<150 kHz
IRS25401S	SO-8	200V	+500 / -700 mA	<500 µA	<500 kHz
IRS25411S	SO-8	600V	+500 / -700 mA	<500 µA	<500 kHz

Electronics likely to get cheaper, courtesy stronger rupee!

With the new government in power, the stock exchanges are reacting positively and the rupee is regaining lost ground. All this is likely to lead to a drop in the price of electronics. However, companies that had increased the prices of consumer electronics products, when the Indian rupee hit a record low of 68.83 to a dollar, will wait and watch before slashing prices. They will also keep a keen watch on the upcoming Union Budget to get a clearer perspective on the new government's policy. Sunil Nayyar, head of sales, Sony India told the media that the company does not plan to do anything immediately but is 'closely watching' the rupee's movement before working out its future strategy on pricing.

Odisha government approves ESDM road map

Taking a step forward to promote electronics manufacturing in the state, the Odisha government has given the go ahead to the electronic system design and manufacturing (ESDM) road map prepared by the India Electronics & Semiconductor Association (IESA). The government officials stated that the process of implementing the recommendations made in the report will begin soon. IESA estimates that the ESDM sector can generate a cumulative revenue of Rs 188,000 million in the state by 2024. According to IESA, the state needs to pump in investments of Rs 73,400 million in the sector, in three phases, to achieve the projected revenue.

Centre approves electronics projects worth Rs 890 million for MP

Central government has given its approval for electronics manufacturing projects worth Rs 886.5 million in Madhya Pradesh. These projects will come up in Badwai (near Bhopal) and Purva (near Jabalpur) districts of the state.

The electronics projects are likely to provide employment to over 17,000 people in the state, said a release issued by the state government's Information Technology department. The release stated that the cost of the electronics cluster at Badwai is estimated to be Rs 483.4 million, for which the central government will contribute about 50 per cent.

The remaining 50 per cent cost will be borne by the state government through the Madhya Pradesh State Industrial Develop-

ment Corporation (MPSIDC) and the units to be set up under the project. The release stated: "For the project, 50 acres have been allotted where world-class high quality facilities of power, water, road and other infrastructure will be provided. The project will be operational within 2.3 years." The units are expected to have a collective investment of Rs 2600 million, said the state government. "The cluster will provide direct employment to 3000 people and 9000 will get indirect employment," said the release. The cost of the project at Purva is estimated at Rs 403.1 million. "The pattern of cost sharing for this project will be the same as in the Badwai cluster," the release added.

Foxconn India affected by Nokia's tax issues

The tax crisis faced by Nokia India may result in a shut down of Foxconn India's Chennai plant. Reportedly, Foxconn is likely to announce a voluntary retirement scheme for its employees soon. It is one of the main vendors for Nokia India and is largely dependent on the Finnish company for its business. Earlier, 450 employees had been given paid leave and asked to come to work only once every week. These employees though were being paid in full, barring incentives on attendance, bonuses and production incentives. The Chennai plant currently operates as a manufacturing unit for Microsoft after its acquisition of Nokia. Foxconn, which makes mobile phone panels for Nokia India, had set up shop in the Special Economic Zone near Chennai.

SOLAR

Piyush Goyal is new energy minister

Piyush Goyal has been appointed as the new minister for power, coal and new and renewable energy. He is expected to pay special attention to solar energy. The ministry of new and renewable energy (MNRE) currently oversees the Solar the Solar Energy Corporation of India (SECI), which is responsible for directing the Jawaharlal Nehru National Solar Mission (JNNSM).

Karnataka cabinet approves revised solar policy

The Karnataka Cabinet has given a clearance to the revised solar energy policy for the state. According to the policy, the farmers will be encouraged to set up solar power plants to produce solar energy up to a capacity of 1 to 3 MW and further sell it to electricity companies. The solar power produced by the farmers will be purchased by the companies at a tariff fixed by the Karnataka Energy Regulatory Commission. The policy also provides for giving financial support to the farmers to install solar plants on their private land, and has the provision for farmers to rent farmland to investors for solar power generation at a rate fixed by the government.

MNRE plans information centre to promote solar

Ministry for New and Renewable Energy plans to set up an 'information centre', which will act as a public awareness wing at the National Institute of Solar Energy's research facility in Gurgaon. The R&D hub for solar technology will now be transformed into an information centre for the general public. The upcoming information centre will work on popularising the advances made in the field of solar power. As per a report, the idea behind the centre is to give people an opportunity to see how a 1 MW solar plant looks like and how it functions. Practical sessions and training programmes will also be held for students.

Indian solar industry did well in April

According to MNRE, solar PV plants commissioned under the National Solar Mission (NSM) have shown consistently high performance in April 2014, even better than the previous month's data. It must be noted that MNRE had already forecasted that capacity utilisation factors (CUFs/CFs) of Indian PV would be promising in the months of April and May.

Snippets

Renesas' new subsidiary in India

Renesas Electronics Singapore recently spun off its India operations into a new subsidiary, for which Sunil Dhar has been appointed as the managing director. "Since 2011, we have been operating via our branch offices in Bengaluru, Delhi and Mumbai. In the last three years, Renesas has gained significant traction in the Indian market. Setting up a wholly-owned subsidiary will allow us to expand our footprint to accelerate business growth." says Dhar.

Philips set to re-enter the Indian mobile handset market

Philips is all set to re-enter the Indian mobile handset market with the launch of a new series of mobile phones in New Delhi. The range includes both smartphones and feature phones. Reportedly, this is a joint venture between Royal Dutch Philips Electronics and Shenzhen SED Industry Co.

Pace Power acquires GE Power Electronics

Pace Power Systems has reportedly acquired GE (General Electric) Power Electronics India's entire business. GE has been providing energy management solutions to telecom operators and telecom tower companies. The deal gives Pace Power the right to use the 'Lineage Power' brand owned by GE.

Karnataka's largest solar power plant commissioned

Welspun Renewable Energy has commissioned the largest solar power project in Karnataka. It was a twin project with a total capacity of 19 MW. The first part of the project, of 8 MW capacity, was commissioned last year, while the remaining 11 MW took just three months to complete.

Punjab to lease vacant panchayat land for solar power

Punjab state government has paved the way for leasing out vacant panchayat land to solar power generators. A majority of solar generators in the state, who have inked the power purchase agreement (PPA) with the state power utility, could not acquire land because of the exorbitant land prices. The government had allocated solar power projects to as many as 26 developers for a total capacity of 250 MW.

UP SMEs to take the solar path

Small and medium enterprises in Uttar Pradesh state are looking to harness the power of solar energy. The initiative is an attempt to deal with the regular power cuts and rising electricity tariffs. Along with the other benefits that come with solar power installations, the 30 per cent central subsidy is a further attraction for the SMEs.

Rolta joins hands with BEL to bid for defence project

Rolta India has joined a consortium alongside Bharat Electronics to bid for a defence project worth over Rs 500 billion. According to reports, Bharat Electronics will be providing the communication equipment apart from other things, while Rolta will be responsible for the software side of the project known as Make India.

Indian e-waste market to grow at 30 per cent during 2014-19

A recently published report by TechSci Research titled "India E-Waste Management Market Forecast & Opportunities, 2019 revealed that the country's e-waste market is expected to grow at a CAGR of around 30.6 per cent during 2014-19. The report stated that the southern and western regions are the largest contributors to the country's e-waste market.

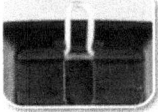
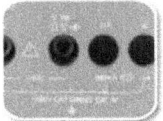

Some major investments in the Indian semiconductor industry

Initiatives taken by the Government of India to promote the Indian semiconductor industry are expected to spur the market in the coming years. The future of the Indian semiconductor design market is dependent on the emergence and success of fabless design companies in India, which would be focusing on creating products for the Indian market. The absence of a legacy has created a unique opportunity for India to leapfrog technology, and India can easily take the lead in electronic systems design and electronic product manufacturing.

India currently contributes less than 5 per cent of the revenues of the global semiconductor market. This share is expected to reach 11 per cent by the end of 2015.

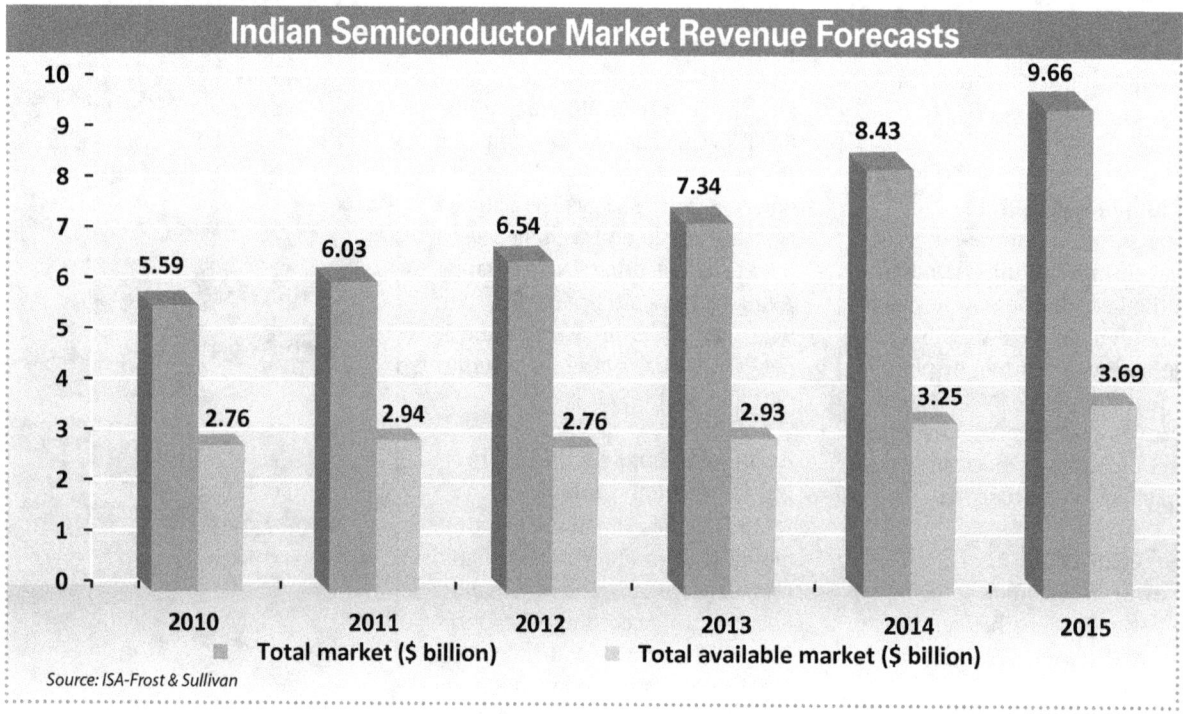

Indian Semiconductor Market Revenue Forecasts

■ Total market ($ billion)   ■ Total available market ($ billion)

Source: ISA-Frost & Sullivan

Some of the major investments in the Indian semiconductor industry include:

- A consortium led by JP Associates, which includes IBM Microelectronics and the systems integrator Tower Jazz, has made an outlay of about Rs 263,000 million for the fab, as per DeiTY
- A consortium led by Hindustan Semiconductor Manufacturing Corporation (HSMC) along with France-based ST Microelectronics and Silterra (Malaysia), has committed to invest about Rs 252,500 million
- Cadence Design Systems has announced that it has completed the acquisition of Cosmic Circuits Private Ltd, a Bengaluru-based provider of analogue and mixed signal intellectual property (IP) cores
- Hitachi High-Technologies Corp has established Hitachi High-Technologies India Pvt Ltd (Hitachi High-Tech India) in Gurgaon to demonstrate its global sales and global sourcing capabilities in India
- Atrenta Inc has opened a research and development (R&D) facility in Noida. The new facility houses Atrenta's current staff and meets near-term growth requirements, keeping pace with the company's global expansion plans
- Forte Design Systems has partnered with CircuitSutra to provide VLSI design services throughout India. Forte and CircuitSutra will co-develop ARM AMBA AXI and OCP-IP models compatible with Forte's Cynthesizer SystemC high-level synthesis (HLS)

Source: Media reports, IESA, Ministry of Information Technology, and press releases

 **Agilent Technologies**

Agilent's Electronic
Measurement Group

Keysight
Technologies

Agilent's Electronic Measurement Group,
including its 9,500 employees and 12,000
products, is becoming **Keysight Technologies.**

Learn more at **www.keysight.com**

Toll-free: +1-800-112-929

Find our products at:

   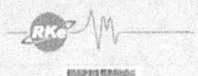 

  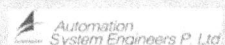

Lean Manufacturing Competitiveness Scheme offers 80% subsidy to MSMEs

The scheme will be implemented in 500 mini clusters across the country during the 12th Plan period

By Srabani Sen

To give an impetus to the ambitious Lean Manufacturing Competitiveness Scheme that was designed to promote the competitiveness of micro, small and medium enterprises (MSMEs) in the electronics sector, the government recently appointed the Quality Council of India (QCI) and the National Accreditation Board for Education and Training (NABET) as the national monitoring and implementing units (NMIU).

Launched in 2009 by the Ministry of Micro, Small and Medium Enterprises, the scheme seems to have lost its sheen over the past five years. But the government cannot succeed in boosting manufacturing unless the competitiveness of the MSMEs is also enhanced, as they form the backbone of the electronics manufacturing sector in India, contributing 40 per cent of domestic manufacturing, besides a substantial contribution to exports.

After it was launched in 100 pilot clusters, the scheme will now be implemented in 500 other mini clusters across the country during the 12th Plan period (2012-17). A group of six to ten MSMEs can participate in the scheme and form a mini cluster.

Lean manufacturing concept

The systems in lean manufacturing aim to identify and remove waste from the manufacturing process, continuously. "Across all segments of the electronics industry, MSMEs have begun to appreciate the benefits of lean manufacturing and are transforming themselves. Many have already embarked on a lean pro-

gramme or are seriously considering one," says an official of the National Productivity Council (NPC).

Subsidy on consultant fees

The Lean Manufacturing Competitiveness Scheme aims to help MSMEs improve the quality of their products and systems and at the same time lower costs, all of which are essential in order to compete in national and international markets. "While big enterprises have been adopting lean programmes to remain competitive, MSMEs have generally stayed away from such programmes due to the cost factor. Besides, experienced and effective lean manufacturing counsellors or consultants are not easily available and are expensive to engage," says an official in the office of the development commissioner, Ministry of MSME.

However, many MSMEs remain unaware that the Lean Manufacturing Competitiveness Scheme offers a subsidy of up to 80 per cent of the consultant fees for each mini cluster. The remaining 20 per cent can be shared by the participating units of the cluster.

Structure of scheme

A three-tiered structure has been envisaged in the scheme. Each mini cluster will work along with the lean manufacturing consultant to implement specific lean techniques in the cluster. The National Monitoring and Implementation Unit (NMIU) forms the second tier, which will be responsible for facilitation, implementation and monitoring of the scheme. At the highest level is the screening and steering committee to provide overall direction to the scheme.

Lean manufacturing consultants will assess the existing manufacturing system of the member units of the mini cluster(s), and formulate detailed procedures and schedules for implementing and achieving lean techniques. A special purpose vehicle (SPV) will be formed in each cluster.

Under the scheme, initially, awareness programmes will be organised in different industry clusters to explain the benefits of lean manufacturing. Thereafter, competent consultants will be deployed to identify cluster-specific needs and suggest appropriate lean techniques to resolve them. The broad techniques envisaged under the scheme include total productive maintenance (TPM), 5S, visual control, standard operation procedures, just in time (JIT), kanban system, cellular layout, etc.

Who can apply

Interested industry associations, as well as groups of about 10 MSME units that qualify under MSME-Development Act, 2006, can submit applications to the National Productivity Council (NPC). ▣

For the guidelines of the scheme, visit www.dcmsme.gov.in

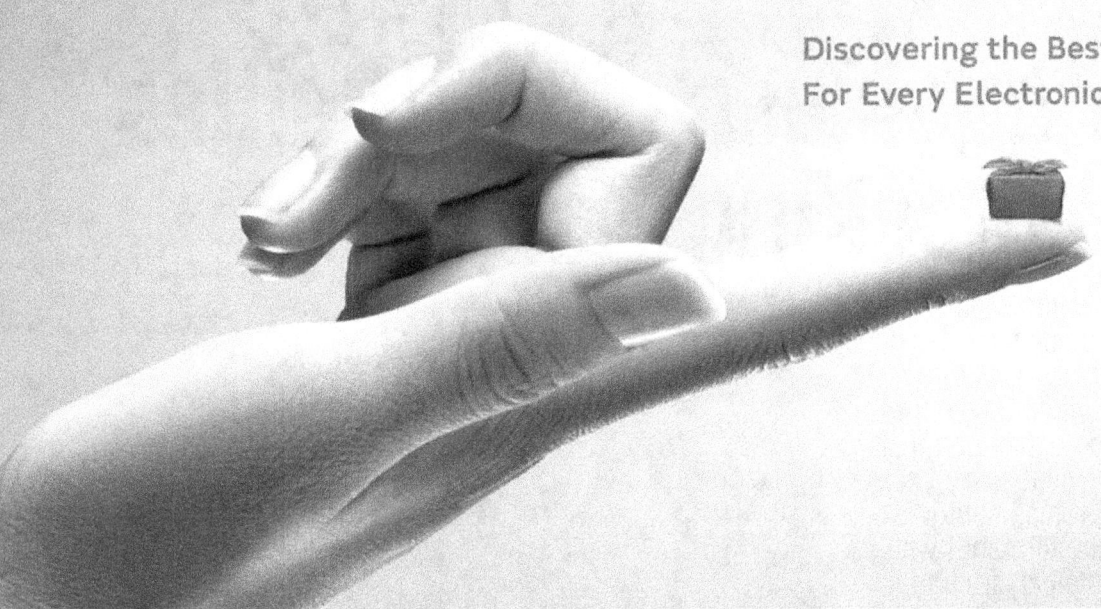

DEFENCE ELECTRONICS
Major business opportunities come knocking

With the Indian government stressing on the need for indigenous design, development and manufacture of defence equipment, and with the country's defence outlay going up by around 10 per cent every year, the defence electronics sector is undergoing a paradigm shift with the entry of private players into a traditionally 'public sector only' domain. Yet, a major part of the market is yet to be explored

By Srabani Sen with inputs from Kartiki Negi

With the 12th Plan's (2012-17) outlay being pegged at over Rs 1000 billion, it appears that the defence/strategic electronics sector in India has the potential to become one of the larger sectors in India over the next 10 years. It is also estimated to grow at an average compound rate (CAGR) of 20-30 per cent. The sector accounts for around 6-7 per cent of the overall Indian electronics market.

The Interim Union Budget 2014-15 allocated Rs 2240 billion for the nation's defence, which is a 9.98 per cent increase over the 2013-14 defence budget. With this annual increase, India is going to be one of the top 10 global spenders on defence.

Moreover, the government is now planning to increase the FDI in the defence sector up to 100 per cent. A Cabinet note said that a foreign company can even take over a domestic entity provided it brings in state-of-the art technology. "This will hugely help in reducing import bill for defence equipment, and help in boosting manufacturing and creating jobs," the note stated.

Although the government is the sole buyer and market marker for defence electronics, and the sector is entirely governed by the defence policy, contrary to general perception, defence electronics has already undergone a paradigm shift due to the participation of private players along with public sector units (PSUs), even though a major part of the market is yet to be explored in the country.

DOES INDIA HAVE THE MANUFACTURING CAPABILITY?

As the world's largest arms buyer, India increased its arms imports by 111 per cent over the past five years and currently accounts for 14 per cent of the world's arms imports. "Strategic electronics sector cannot depend on imports. It requires a very high level of quality in terms of input materials, production processes as well as end testing. If India increases its strategic electronics production, it will be a good exposure for domestic manufacturers," says N Ramachandran, managing director, MEL Systems and Services Ltd.

Alternative to import dependence: Manufacturing of defence products in India was opened up to private and foreign players in 2001. Till then, only defence PSUs like Hindustan Aeronautics (HAL), Bharat Electronics Limited (BEL), Bharat Dynamics Limited (BDL), etc, manufactured defence equipment in the country. These companies currently account for about 20 per cent of the Indian defence equipment market. But with the introduction of the defence procurement procedure (DPP) in 2002, which was further amended in 2006, a handful of private companies like Tata Motors, L&T and Mahindra Aerospace have forayed into this sector. This is a paradigm shift in the area of strategic electronics production in India. However, the private sector is estimated to have a relatively small share—just 5 per cent--of the defence equipment market. So the major part of India's needs, over 70 per cent of defence equipment, is still met through imports.

India's total imports of defence equipment amounted to a mammoth Rs 400 billion in 2013, and this huge market can now be tapped by the private companies also. The Indian government has stressed on the need for indigenisation in design, development and production of defence equipment in the Defence Procurement Policy announced in 2013. Between 2004 and 2013, the government issued 209 licences to Indian companies to manufacture defence equipment domestically. The pace of granting licences has also picked up over the last five years.

Indian manufacturing capability: It is quite clear that 70 per cent of the demand, which is now being met through imports, can be tapped by the private manufacturers, though currently, only 5 per cent is being catered to by them. So, do these private companies lack the capability? According to industry insiders, India definitely has the manufacturing capabilities and the infrastructure for electronics, mechanical or optoelectronics. But unfortunately, right now, we lack the technology and design capability for strategic electronics. So we need to borrow the technology. Even the technology that India has transferred is outdated.

Says Vedaprakash G, general manager, business development, Centum Electronics Ltd, " Centum has made significant

T Vasu, director, Tandon Group

N Ramachandran, managing director, MEL Systems and Services Ltd

Col K V Kuber (Retd), advisor, aerospace, defence and homeland security, NSIC

investments to establish design and manufacturing capabilities for the defence and aerospace segments since 2006 and we have begun to see the benefit through offsets contracts with international defence companies as well as domestic strategic customers. We now have a strong capability to manufacture various complex electronic subsystems here in India. Going forward, Indian companies will need to evolve by continuing to invest in design and developing state of the art products. There are only few good Tier II players in this segment." Centum, a leading globally renowned electronics company, develops and manufactures high reliability electronic products such as missile guidance subsystems, on board computers, power subsystems, launch vehicle subsystems, flight electronics and frequency control products to the stringent requirements of defence and ISRO.

Adding to this, Rajoo Goel, secretary general, ELCINA, says, "In our country, there is sufficient indigenous capability to manufacture superstructures, containers, hydraulics, electromechanical assemblies, etc. However, we lack the capabilities required for the core electronic segment of combat systems, where we are still dependent on foreign manufacturers. While noticeable growth has been seen in Tier I and III vendors, we have been lagging behind in exploiting opportunities in the Tier II category, which contributes significantly to the overall cost of strategic systems."

However T Vasu, director, Tandon Group, states, "Today the Ministry of Defence (MoD) aims to create conditions conducive for domestic manufacturers, both public and private, to play an active role in this domain. It now aims at expediting decision making, as well as simplifying contractual and financial provisions to establish a level playing field

DEFENCE ELECTRONICS: WHAT INDIA CAN MANUFACTURE

India has the capability to manufacture the following products:
- Electronic warfare equipment
- Satellite communication and homeland security solutions
- Electrical systems that comprise components like fuel systems, fire prevention systems, connectors and electrical wiring, etc
- Instrumentation and display systems comprising components like fly-by-wire and autopilot systems, aircraft data display systems, G-load, side slip sensors, temperature and pressure sensors, etc
- Radio and communication systems
- Tank electronics • Electronics for missiles
- Avionics systems • Radar systems
- Underwater electronics
- Communication equipment
- Tools, testers and ground equipment

Source: N Ramachandran, managing director, MEL Systems and Services Ltd

Gautam Awasthi, general manager, marketing, Agilent Technologies

Dr Ajay Kumar, joint secretary, Department of Electronics and IT (DeitY), govt of India

Pavan G Ranga, CEO, Rangsons Electronics Pvt Ltd

for the Indian companies. However the implementation is not absolutely industry friendly. There should be more of a win-win situation than the existing one."

Rise in domestic manufacturing: Despite all the limitations, strategic electronics manufacturing in India has been growing steadily. From Rs 57 billion in 2007-08, it was projected to touch Rs 120 billion by the end of 2013 (confirmed figure still not available) and exceed Rs 140 billion during 2013-14. This high potential emerged particularly after the revised Offset Policy was announced in 2012, offering greater opportunities to indigenous electronics manufacturers. The Offset Policy allows companies to supply strategic electronics equipment in collaboration with MNCs, on the condition that sufficient value addition is done locally. The new Offset Policy mandates foreign defence equipment vendors to source 30 per cent of the contract value, locally, if their order value exceeds Rs 30 billion. This should definitely spur domestic production of strategic electronics.

However, not many can venture into the strategic electronics domain, as the MoD has strict rules relating to procurement, defined under the defence procurement procedure (DPP) introduced in December 2002. The DPP, which was amended in 2006, mandates that the suppliers of defence electronics products or equipment have to sign pre-contract integrity pacts with the government.

Meagre contribution from MSMEs: The amended DPP also was to act as a catalyst to enhance the potential for micro, small and medium enterprises (MSMEs) to be part of the indigenisation drive, as well as help in broadening the defence R&D base of the country. Yet, the contribution of the MSMEs is meagre. Says Col K V Kuber (Retd), advisor, aerospace, defence and homeland security, NSIC, "MSMEs are not yet capable of making a tank, a gun or an aircraft. They can only do less critical tasks such as fabrication, painting, limited machining or making small parts. So talk of direct participation of MSMEs in the defence sector is far fetched. To contribute significantly in this sector, MSMEs have to work as a group, like one firm doing the machining, another fabrication, and yet another assembling the product. Together, they can develop the complete product." In fact, BEL's supply chain includes 300-400 MSMEs.

Also, defence contracts take a long time to execute—maybe even two years. It becomes difficult for an MSME to wait that long. MSMEs are already doing sub-systems for PSUs.

T Vasu is of the opinion that the policies should be made conducive for the MSMEs to participate as a large number of them can contribute to the defence requirements, particularly in manufacturing. "The strict 'no cost, no commitment' (NCNC) terms of the policy, kills (NCNC) in the policies kills the motivation of the MSMEs. In many cases the expense involved in proof of concept, demos and field trials simply go waste and further discourages the companies from participating in defence business," he adds.

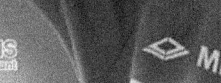

HOW SES SUPPORTS DEFENCE ELECTRONICS SECTOR

The idea of a Strategic Electronics Summit (SES) was mooted by ELCINA five years ago. It has grown to become a platform that brings together all the stakeholders of the strategic electronics sector to explore untapped opportunities. "The Defence Offset Policy and DPP have opened up a host of opportunities which the domestic industry must seize. The objective of SES is to highlight these opportunities to all industry players keen to gain from and contribute to the growth of the strategic electronics industry in India, and enable greater domestic value addition and indigenisation," says Rajoo Goel, secretary general, ELCINA.

The fifth edition of the SES will be held from July 30-31, 2014 in Bengaluru. "The production of strategic electronics in India has been growing steadily year on year but is still insufficient to meet local demand. It can safely be said that the value of the electronics alone will rise faster than the total value of the strategic equipment and thus become more and more important, strategically as well as commercially," says Rajoo Goel.

Objectives of the SES: SES aims to bring all stakeholders on to one platform to enable better communication and to understand the defence establishment's requirements. Its vision is to create indigenous capabilities to manufacture defence equipment; hence, it makes an effort to highlight new policy initiatives. One of the key objectives of the SES is to involve the MSMEs, and encourage them to meet the requirements of the defence sector.

Fifth edition of ELCINA SES: "The fifth edition of the SES will be of great relevance as, in the last six months, the mood of the government has dramatically changed regarding its role in boosting domestic manufacturing," says Rajoo Goel.

"Government wants to focus on 'made in India' products. To transform this thought into action, the SES will play a major role—by providing a platform for the stakeholders to meet and debate over ideas on how to take the sector forward. This exhibition also showcases the capabilities of the existing units, and will inform people about what's new in the market," says N Ramachandran, MD, MEL Systems and Services Ltd.

SES also has NSIC (National Small Industries Corporation) as a participant; hence, many micro, small and medium establishments will display their products at this event. "MSMEs might not make too many strategic electronics parts at the moment, but they are very much into electronics. So, if they can get the right business opportunities, they can contribute more to this sector. SES is a good platform to provide them these opportunities," says Col K V Kuber (Retd), advisor, aerospace, defence and homeland security, NSIC.

For more details on the event, visit www.elcina.com

IMMENSE BUSINESS OPPORTUNITIES

An increased budgetary allocation for the armed forces, the army's plans to replace old equipment, and compulsory sourcing of Indian components by foreign defence equipment vendors are some of the key drivers for the growth of India's strategic electronics industry, which holds immense business opportunities.

Over the next 10 years, the government is planning on acquiring defence electronics worth US$ 150 billion. Recently, addressing the 5th edition of the Tamil Nadu Manufacturing Summit, organised by the Confederation of Indian Industry (CII), A Sivathanu Pillai, chief controller (R&D), Defence Research and Development Organisation (DRDO), said, "The requirements of the defence sector are opening up opportunities worth US$ 150 billion for the industry, and out of this, over $ 100 billion is open to Indian companies. The figure has been arrived at based on the main orders expected by DRDO during the period mentioned," he said. DRDO is developing products for orders worth Rs 160 billion, and many of its projects are in the completion stage. Hence, it is looking for companies that have the capability to supply to the organisation.

Huge market potential: Observing the huge potential ahead, domestic as well as global private sector companies are making inroads into the strategic electronics market. Of late, domestic companies that made only small parts have started developing in-house capabilities. While a large number of foreign companies are present in this space, Indian companies like Tata Motors, Larsen & Toubro (L&T), Bharat Forge and Ashok Leyland are also making their presence felt in this domain.

Says N Ramachandran, "This domain has enormous potential for manufacturers because, over the last six months, the government has made a commitment that it would like to have more and more defence production happen in India. Many new tenders have been announced to give primary importance to Indian manufacturers." The government is committed to cutting down on imports for two reasons—first, it is expensive to import; and second, the products and systems are really strategic in nature.

It is not possible for the PSUs to cater to this huge demand. Moreover, many of them do not even have all the capabilities to do so. For instance, HAL doesn't have design capabilities and does licensed production. On the other hand, only HAL has aerospace capabilities. This creates a lot of space for the private companies to step into.

Those who enter this domain will surely be around for 20 to 25 years. That's the kind of sustainability to be expected. This sector can ensure profitable business for companies that are into design, custom-built products, contract manufacturing, component manufacturing, systems integration, etc.

Contracting: This is a big opportunity for private firms and OEMs. For example, Tata Power Strategic Engineering Division (SED) is a prime contractor of the Ministry of Defence (MoD) for indigenous defence electronics production. In 2011, it bagged a Rs 9500 million contract from the Indian Army to manufacture two electronic warfare systems. The

DEFENCE ELECTRONICS: WHAT CAPABILITIES INDIA HAS

Strategic electronics companies have strengths in:

- Well-defined manufacturing processes–ISO 9001, ISO 14001, etc • BS7799 certification for IP security
- In-house capabilities for industrial design
- Design and development in the mechanical, electrical and software space • Prototyping • Testing and certification
- Strong tie-ups with leading component suppliers
- 30 years of manufacturing experience
- Sensitive to customer time lines, quality requirements and cost targets

Manufacturing offerings include

- Contract manufacturing • Electro-mechanical parts and assemblies • Sheet metal parts • Plastic parts
- Machined parts • Sourcing of systems and components
- PCBs, chips, discrete electronic components, switches, connectors, cables and harnesses • Castings and forgings
- Packing materials • Assembly operations
- PCB assemblies • Cabling, routing and wiring
- Building complete assemblies and systems
- EMS manufacturing • Defence electronics and electro-mechanical products • Various testing capabilities

Source: Research Paper on 'Opportunities Landscape in Strategic Electronics' prepared by ELCINA and IESA

Rajoo Goel, secretary general, ELCINA

Vedaprakash G, general manager, business development, Centum Electronics Ltd

company won another Rs 11,700 million contract to modernise 30 Indian Air Force bases. It has the ability to design, develop, manufacture, assemble and upgrade mission-critical systems. In 2012, Bengaluru-based Centum Group took up a contracting order to supply to Europe-based defence solutions provider, Thales and added a few more international strategic customers like Rafael, L3 and others. Bharat Forge is also aiming to become a major player in the artillery and specialised vehicles segment. Bengaluru-based Dynamatic Technologies makes assemblies of vertical fins for the Sukhoi 30 Mk-I fighter bombers.

Several other small companies like Avasarala Technologies, DefSys, Ravilla and Taneja Aerospace have acquired advanced technological capabilities and will be able to upgrade from making parts to developing entire systems.

Three types of contracts: In defence contracts, there are three types of manufacturing. In 'build-to-print', the contractor manufactures according to designs provided by a client. While in 2010, 92 per cent of the private

sector's orders that came to India were for build-to-print, this figure has dropped to 80 per cent in 2014.

In the second variety—'build to specifications'—the project executor owns the intellectual property. Here, the client provides the idea of a product, and the executor manufactures based on that. Often, these products require developing new intellectual property that may belong to the Indian contractor.

The third type of manufacturing involves developing products from scratch, either to create new capabilities for a particular defence need or to upgrade an existing product.

"Indian manufacturers have now started building capabilities in the second and third categories. The Indian private sector has taken big strides forward, technologically, in the last few years and can now match international standards," says Pavan G Ranga, CEO, Rangsons Electronics Pvt Ltd.

Maintenance, repair and overhaul: Maintenance, repair and overhaul (MRO) for aerospace and defence equipment is a big opportunity for private firms and OEMs. Most major equipment remains in service for two to three decades. MRO facilities can be established in the private sector or with OEM participation. However, there is a growing awareness within the Indian defence establishment regarding the value of outsourcing non-core maintenance activities to third party operators. Major global aviation industry firms are already eyeing the market in India. The revised DPP provides the opportunity to establish public-private partnerships for MRO.

Despite the exponential growth projected for the MRO business in India, the country has still not sorted out the certification issues related to training the required workforce. The certification, given by the Directorate General of Civil Aviation (DGCA), is not yet comparable with mandatory European standards. The DGCA is now in the process of making its certification equivalent to the European Aviation Safety Agency (EASA), which is mandatory for MROs that hope to service the international airlines.

Offset partnership: As per the terms of the DPP 2006, there is an option to offset up to 30 per cent of the total foreign exchange the government spends. It involves the foreign vendor being contractually obliged to export products (in order for India to recover a percentage of foreign exchange), components and services; make investments in industrial infrastructure for services, co-development, joint ventures and co-production of defence products and components.

Domestic companies can target the offset areas where there are immense opportunities. OEMs from US and EU are proactively looking for Indian offset partners (IOPs) with technological capabilities, capacity and IPR process maturity. Additionally, business opportunities through offset for systems design, engineering and testing services can also be targeted by domestic companies. As an example, an induction of 300 helicopters will provide offset opportunities, apart from subsequent MRO requirements.

From 70 MHz to 4 GHz:
Powerful oscilloscopes from the T&M expert.

Fast operation, easy to use, precise measurements –
That's Rohde & Schwarz oscilloscopes.

R&S®RTO: Analyze faster. See more. (Bandwidths: 600 MHz to 4 GHz)
R&S®RTE: Easy. Powerful. (Bandwidths: 200 MHz to 1 GHz)
R&S®RTM: Turn on. Measure. (Bandwidths: 350 MHz and 500 MHz)
HMO3000: Your everyday scope. (Bandwidths: 300 MHz to 500 MHz)
HMO Compact: Great Value. (Bandwidths: 70 MHz to 200 MHz)

All Rohde & Schwarz oscilloscopes incorporate time domain, logic,
protocol and frequency analysis in a single device.

Take the dive at www.scope-of-the-art.com/ad/all
Email: sales.rsindia@rohde-schwarz.com

Meet us at:
Strategic Electronic Summit, SES
for Defence & Aerospace
Booth No.-B1
30 - 31 July 2014, BIEC, Bengaluru

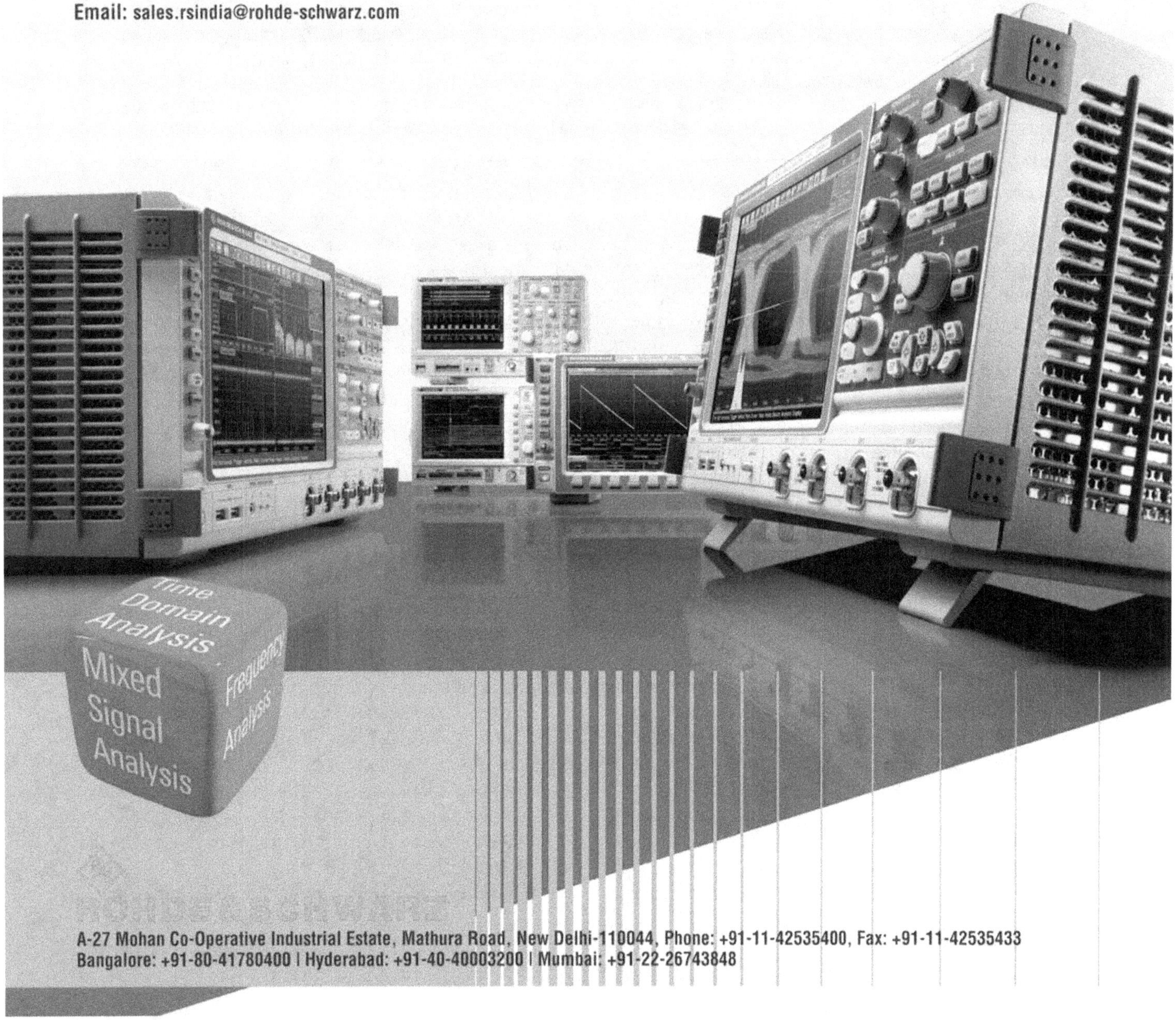

A-27 Mohan Co-Operative Industrial Estate, Mathura Road, New Delhi-110044, Phone: +91-11-42535400, Fax: +91-11-42535433
Bangalore: +91-80-41780400 | Hyderabad: +91-40-40003200 | Mumbai: +91-22-26743848

TURNING CHALLENGES INTO OPPORTUNITIES, DRDO HAS DEVELOPED INDIGENOUS AND INNOVATIVE TECHNOLOGIES FOR INDIAN DEFENCE SECTOR

By Ravi Kumar Gupta, scientist G & director, directorate of public interface, DRDO

Electronics is a key area of defence technologies, and is a very important part of nearly all the weapon systems, platforms and equipment designed and developed by the Defence Research and Development Organisation (DRDO) for the Indian Armed Forces. DRDO translates the products it designs and develops into manufactured items that are used by the defence forces, with the help of manufacturers in the public (ordnance factories, DPSUs such as BEL, ECIL, etc) and the private sectors.

DRDO has close associations with over 800 industries that include a large number of SMEs. DRDO and the Indian defence sector had to pass through a tough phase for several decades due to the lack of a strong industrial base in the country, inadequate testing and infrastructure facilities, and the shortage of trained/skilled human resources. The situation was further complicated with the sanctions imposed by the West that followed Pokhran I and Pokhran II, as well as the launch of the indigenous ballistic missiles.

Turning these challenges into opportunities, we went on to develop indigenous and innovative technologies. Today, we have achieved a very high degree of self-reliance in many crucial areas such as radars, sonars, navigation systems and sensors, high performance computing (hundreds of teraflops), among others. These developments have also led to a quantum leap in the growth of Indian industries. Today, the production value of the electronics systems designed and developed by DRDO, which are being manufactured by the Indian electronics industry, amounts to more than Rs 250 billion. And this figure is rising rapidly. Thus, the defence electronics sector in the country, led by DRDO, is poised for rapid growth with the active participation of Indian industry.

DRDO has 11 laboratories engaged in R&D, which are related to various facets of electronics and its applications. Thus, from the design and development of solid state materials, devices, microwave tubes, microprocessors, communication systems, command and control systems, radars, electronic warfare systems and electro-optical systems, to advanced computing, artificial intelligence and robotics—DRDO's R&D activities cover it all.

In addition, about a dozen other DRDO laboratories—which are developing systems and applications for aerospace, combat vehicles, combat engineering and the navy—have highly qualified groups working in specific areas of electronics related to their requirements.

System and sub-system integrators: Systems integration with outsourced components and sub-systems is the most viable business option in this segment. All major defence contractors are usually integrators who procure components, sub-assemblies and sub-systems from the MSMEs to configure defence systems as per the parameters prescribed by the buyers. For example, India's ambitious project F-INSAS (Future Infantry Soldier as a System) entails integration of multiple technologies. Hence, vast opportunities exist for domestic as well as foreign MSMEs.

DRDO has been at the forefront of missile development with its Integrated Guided Missile Development Programme. Some of these missiles may be co-produced in India by the public sector. Hence, there is an opportunity for the private sector to indigenise sub-systems or manufacture them with the transfer of technology (ToT).

However, Pavan G Ranga says, "Systems integration no doubt is a good business opportunity, but more important is the core engineering and core technologies that need to be developed in the country. Moreover, the integrators need to add value; otherwise it is not a sustainable business model." He is of the opinion that sub-system manufacturers have the biggest business opportunities. "About 50 per cent of the manufacturing opportunities will comprise components and sub-systems manufacturing. It is a huge opportunity that players are looking at," he says.

T&M device suppliers: The increase in indigenous design and manufacturing of aerospace and defence systems has resulted in a growing demand for high performance, state-of-the-art test and measurement (T&M) equipment. There is a demand for commercial off-the-shelf (COTS) equipment that satisfies military (MIL) standards in the strategic electronics segment.

"T&M equipment and solutions for aerospace and defence are based on advanced technology for use in radar installations, electronic warfare, military communications, satellites, guidance, avionics, intelligence, surveillance and beyond," says Gautam Awasthi, general manager, marketing, Agilent Technologies (now Keysight Technologies).

Other opportunities: Besides the above opportunities, there is a US$ 50 billion opportunity in the electronics system design and silicon fabrication space requiring high end technology transfers in the next 10 years. Homeland security, environmental technologies and training services are another big slice of the pie that is estimated to be worth US$ 2.5-10 billion, within a similar time frame.

Defence projects open up opportunities: Currently, there are a number of defence projects that will lead to vast business opportunities. For example, the Tactical Communication System (TCS) for the Indian Army, which has been in the planning stage for over a decade, is an Internet protocol-based mobile system. This programme will truly open up new opportunities to the private sector.

FINSAS, a multi-billion-dollar project, is being steered by DG Infantry and DRDO, along with private sector partners actively involved in indigenous development.

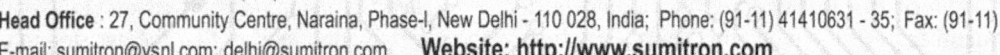

The Indian Navy's modernisation plans include the acquisition of naval weapons, which would require radars, sonars, sensors, etc. Most of these will be provided by BEL, with private sector participation.

CHALLENGES AND ROADBLOCKS

Although sustainability is very high in this domain, breaking into this industry is very tough, as the entry barriers are extremely high. At a seminar on strategic electronics, Dr Ajay Kumar, joint secretary, Department of Electronics and IT (DeitY), Government of India, had said, "Companies in the defence sector require patience, perseverance and determination to make it. But all those who have made it, unanimously believe that having attained the ability to participate in defence tenders has been a good experience."

Difficult procurement procedures: The procurement procedures are very difficult to crack as there are a number of agencies that deal with just one procurement. "It is easier to export defence components, as there are some impediments that Indian companies face when it comes to participating in MoD tenders," says Col K V Kuber.

"The government should make the whole process more transparent, and announce what equipment they would need in the next 10 years. They should make this public," says N Ramachandran.

Huge capital expenditure: Venturing into the strategic electronics industry requires huge capital expenditure in the initial years, which is a constraint for startups and small companies. Once a company starts doing well in this domain, it can foray into the export market as well.

Lack of cutting-edge technology: Cutting-edge technology is required for defence products, which is not available in India. Hence, a lot of technology has to be acquired from abroad, which is expensive. The acquisition process requires heavy investments and a long waiting period. "Companies need to have deep pockets to do business in the defence sector," says N Ramachandran.

Slow tender process: The tender cycle in Indian defence contracts sometimes takes too long for closure— up to 18-24 months. In order to encourage private players' active participation in defence opportunities, this should be kept to within a reasonable 'start-end' period. "A company has to invest in technology, infrastructure and then bid in projects. Projects are decided on the basis of the lowest cost quoted. If a firm does not get that particular business, it has to wait for another cycle," states N Ramachandran.

Need for testing infrastructure: There isn't a good test infrastructure for strategic electronics, barring the labs within government premises such as at BEL, which are always occupied. The government should set up a number of testing centres, feel industry experts.

Policy raises confusion: Although the industry finds the offset policy to be an excellent initiative by the government, they feel that it needs to be made simple enough for all to understand. For example, many in the industry are unable to participate in the process because of the confusion over issues like whether or not they require an industrial licence.

Critical tax issues: Tax issues are extremely critical for defence products. If a vendor imports equipment, it does not pay any duty but if it buys from a domestic manufacturer, it needs to pay tax. So experts feel that all indirect taxes for the defence sector must be eliminated.

INDUSTRY'S RECOMMENDATIONS TO GOVT

Although the government has taken a number of positive steps, many limitations and roadblocks in the procedures remain. "No foreign player would be interested in investing in this sector because of the 26 per cent FDI cap, which should be increased to 45 or 50 per cent," says Vedaprakash G.

The following are some of the recommendations of the industry.

Ecosystem to support indigenous defence products: The Indian government has to pledge to build an ecosystem to support indigenous defence products with a clear roadmap. All ICTE programmes that are non-platform/non-weapons-based need to be opened up for the industry's participation on a competitive basis so that products can be developed locally and sold globally. There should be no PSU reservations.

Clear roadmap for import reduction: Though immediate reduction in imports is quite difficult, channelised imports are important to boost self-reliance in strategic electronics. A clear roadmap towards import reduction by the MoD needs to be published and audited by the CAG, with aggressive targets for the ICTE sector.

Investments in components and composites: It is clear that India needs investments in components and composites. So, wherever such investments come with a control on full intellectual property rights, no 'end-use' restriction should be imposed to ensure more participation of Indian firms in strategic electronics.

Faster decision-making processes: ELCINA had requested the government to ensure that the decision-making process is faster, as MNCs are unable to wait for too long. There are instances when it has taken four-five years from the proposal

PROGRAMMES UNDER THE 12TH PLAN THAT WILL DRIVE GROWTH IN THE DEFENCE ELECTRONICS SECTOR

Programmes that include development and procurement of:
- Tactical Communication Systems (TCS)
- Network Centric Warfare (NCW) systems
- Electronic Warfare (EW) systems
- Future Infantry Soldier As a System (FINSAS)
- Tank electronics • Air defence systems • Avionics
- Navigation equipment • Radar and sonar systems
- Sensors, night vision devices, power generation equipment, and a host of associated and embedded electronics

stage of a project to the contract-signing stage. In this process, a lot of extra costs are incurred due to the time overrun. "While the Offset Policy has opened up opportunities for the Indian industry to participate in defence business with foreign OEMs, the delay in the decision making process is a big hindrance in winning business opportunities," says T Vasu.

Through a research paper on *'Opportunities Landscape in Strategic Electronics'* prepared by ELCINA and IESA (India Electronics and Semiconductor Association), the industry has made certain recommendations to the government.

Short-term recommendations
- To encourage and provide avenues for India-based design/manufacturing houses to compete with foreign players (Taiwanese/Korean/Chinese) through government incentives/preferences, such as waivers in taxes/duties, etc.
- Right to information (RTI) coverage should be provided for business opportunities in defence so that transparency is established while evaluating and choosing a service provider in a tender situation.
- Encouragement should be given to build industry-academia partnerships. These should be structurally covered and R&D initiatives must be compulsory.
- R&D in academic institutions should be encouraged in areas of strategic interest.
- The government should allow 49 per cent FDI in the defence sector to usher in growth in domestic defence production.

Long-term recommendations
- To provide avenues for capability building in blending horizontal services with vertical's business needs

which are currently imported, for example, telematics, navigation systems in automotive industry.
- To send students abroad under subsidy to study in designated strategic areas, and ensure their return and employment in the strategic electronics industry.
- Maintain active contacts with NRIs working in strategic areas and create an environment for them to return to India.
- DPSUs should outsource more and more sub-systems and assemblies to the private sector.
- To create a ToT monitoring cell with competent people from DPSUs, private industries, DRDO and academic institutions, and empower them to effectively monitor and implement the ToT process.

A detailed ToT proposal specifying the technology to be transferred and the step-by-step mode of transfer has to be mandated in all large value purchases.
- To set up a National Defence Design University (NDDU), exclusively for studies and R&D in critical technologies for defence and the design of futuristic weapon platforms.
- Promote FDI in high cost, high technology components such as chip/processor manufacture under strict monitoring control. MoD to provide funding to private companies that are into R&D dedicated to the nation's defence, like how the US Defence Department funds private organisations that are in defence R&D in the US.
- To provide incentives/waivers of taxes to Indian design/manufacturing houses should they set up a manufacturing facility to indigenously manufacture a product supplied by the OEM.

ELCINA's research paper *'Opportunities Landscape in Strategic Electronics'* also pointed out a few other actions that can be actively considered by the strategic electronics industry:
- Creation of infrastructure for manufacturing of components—complex PCBs, SMDs, FPGA-based microprocessors, etc, catering to domestic as well as export markets under strict monitoring control.
- Collaboration with leading OEMs on a ToT/JV basis in niche technology areas.
- Collaboration with academic institutions/sponsoring projects to gain technologies.
- Creation of in-house R&D facilities, either independently or jointly with other vendors.
- At the Tier II and III levels, a few vendors have already been associated with defence production. They need to convert their experience into 100 per cent indigenisation and scale up to reach the Tier I category. Simultaneously, effort must be put into building expertise in newer technologies and venturing into limited production rather than waiting for the current technology to mature and again getting into the loop of importing. **EB**

TOP 20 ELECTRONIC COMPONENTS MANUFACTURERS IN INDIA

By Richa Chakravarty

The recent policy initiatives to boost the electronics industry will hopefully revive the dormant Indian electronic components manufacturing sector in the years to come. The demand for electronic components is estimated to grow at an average yearly rate of about 12.5 per cent over the next three years as the market is estimated to grow to around US$ 15 billion in FY 2013-14.

To find out more about the companies (manufacturers) in this domain, Electronics Bazaar attempted to rank the top electronic components manufacturers in India. Most of these companies have registered a higher revenue than last year, indicating growth in the sector. However, despite maintaining the first position, Epcos India witnessed a slight dip in revenue in FY 2012-13 in comparison to FY 2011-12. While most of the companies retained their earlier rankings, Keltron moved two positions up in the ranking. Comso Ferrites also moved up three rungs.

Most of the companies ranked are exclusively into component manufacturing in India, while some are into multiple businesses in the country. Many companies like PLA, Andhra Electronics, Cedicom, etc, could not be ranked as their revenue figures were not accessible. Some companies also declined to disclose their revenue figures. We have not included companies that manufacture connectors or are solely manufacturing semiconductors. We chose not to include Bharat Electronics Ltd (BEL) in the ranking as it is a multi-product and multi-unit company.

The revenue figures have been procured from the website of the Ministry of Corporate Affairs (MCA), Government of India, and then verified by the companies. Some companies also sent us their balance sheets.

Disclaimer: While Electronics Bazaar's editorial team has taken the utmost care to contact all possible sources to make the list comprehensive, we may have inadvertently left out a few companies from this list. Information contained in the following report has been obtained from sources deemed reliable, and Electronics Bazaar makes no claims with respect to its accuracy. Neither Electronics Bazaar nor its affiliates, officers, directors, employees, owners, representatives nor any of its data or content providers shall be liable for any errors or for any actions taken by any entity on the basis of this information.

TOP 20 ELECTRONIC COMPONENTS MANUFACTURERS IN INDIA
Ranked on the basis of revenue, FY 2012-13

Rank	Company	Revenue (in Rs, million) FY 2012-13	Rank	Company	Revenue (in Rs, million) FY 2012-13
1	Epcos India Pvt Ltd	5738.31	11	General Industrial Controls Pvt Ltd	389.79
2	Vishay Components India Pvt Ltd	2074.40	12	Incap Ltd	353.36
3	Globe Capacitors Ltd	740.52	13	Alcon Electronics Pvt Ltd	323.23
4	Deki Electronics Ltd	665.03	14	Watts Electronics Pvt Ltd	292.67
5	Keltron Component Complex Ltd	611.10	15	Solid State Systems Pvt Ltd	258.76
6	Victor Component Systems Pvt Ltd	550.74	16	Prismatic Engineering Pvt Ltd	137.80
7	Cosmo Ferrites Ltd	494.10	17	Speedofer Components Pvt Ltd	130.00
8	Desai Electronics Pvt Ltd	493.67	18	Thakor Electronics Ltd	114.61
9	BG LI-IN Electricals Ltd	479.03	19	Cermet Resistronics Pvt Ltd	106.47
10	Genius Electrical & Electronics Pvt Ltd	399.76	20	Gujarat Poly AVX Electronics Ltd	94.68

1 Epcos India Pvt Ltd

Annual revenue (2012-13): Rs 5738.31 million

Natarajan Balakrishnan, MD and CEO

Epcos India is a member of TDK-EPC Corporation, which is headquartered in Tokyo, Japan. In India, Epcos develops, manufactures and markets electronic components, modules and systems. In 2013, the company commenced production of transformers at its Kalyani plant, with technical knowhow from TDK, for use in white goods.

- Established in 1995
- Headquarters in Kalyani
- Manufacturing facilities in Nashik, Kalyani and Bawal
- Components manufactured: LV and MV PFC capacitors, soft ferrite cores, etc
- Sectors it caters to: Industrial, infrastructure, IT, consumer, communications, automotive, etc

Contact details: *Ph: +91-80-40390640, sales.in@epcos.com, www.epcos.com*

2 Vishay Components India Pvt Ltd

Annual revenue (2012-13): Rs 2074.40 million

Rajan Shringarpure, MD and director, operations

Vishay Components India Pvt Ltd (VCIPL) is a wholly owned subsidiary of Vishay Intertechnology Inc, a US-based company. The company started its operations in 1958 under the name of Philips; later in 2003, Vishay acquired it. Vishay has two manufacturing facilities in India. It manufactures both passive and active components in India.

- Established in 1958
- Headquarters in Pune
- Manufacturing facilities in Pune and Mumbai
- Components manufactured: Film capacitors, resistors, SCRs, etc
- Sectors it caters to: Industrial, automotive, consumer, telecommunications, defence, etc.

Contact details: *Ph: +91-20-30516200, sougata.ganguli@vishay.com, nirbhay.gopal@vishay.com, pramod.vaidya@vishay.com; www.vishay.com*

3 Globe Capacitors Ltd

Annual revenue (2012-13): Rs 740.52 million

Sanjay Agarwal, MD

Globe Capacitors started its operations with paper/foil-based technology and gradually adapted to manufacturing metallised polypropylene capacitors. Today, the company is equipped with a state-of-the-art fully automatic plant. The company exports 50 per cent of its production. Its products have been certified by UL, CE, ERDA, BIS and TMP.

- Established in 1982
- Headquarters and manufacturing facility in Faridabad
- Components manufactured: AC motor-run capacitors, electrolytic start capacitors, etc
- Sectors it caters to: Home appliances and electronic applications

Contact details: *Ph: +91-129-4275500, info@globecapacitors.com, www.globecapacitors.com*

4 Deki Electronics Ltd

Annual revenue (2012-13): Rs 665.03 million

Vinod Sharma, MD

Deki was established in technical collaboration with Okaya Electric Industries Company Ltd, Japan. It manufactures plastic film capacitors at its state-of-the-art automatic plant with latest and sophisticated machinery imported from Korea, Taiwan, China and Italy. Its current capacity is 1.2 billion pieces per annum.

- Established in 1984
- Headquarters and manufacturing facility in Noida
- Components manufactured: Wide range of capacitors
- Sectors it caters to: Consumer electronics, telecom, lighting, medical, industrial and automotive electronics, etc.

Contact details: *Ph: +91-120-2584687/88, 2585457/58, shariq@dekielectronics.com, www.dekielectronics.com*

5 Keltron Component Complex Ltd

Annual revenue (2012-13): Rs 611.10 million

P S Anandanarayan, MD

Keltron Component Complex Ltd (KCCL) is a government of Kerala undertaking and is a subsidiary of the Kerala State Electronics Development Corporation Ltd. KCCL is aiming for exponential growth by achieving a turnover of Rs 2000 million (US$ 43 million) by 2016. Major expansions are being planned by the company, especially in the areas of screw type capacitors, high voltage radial capacitors, MPP capacitors and power factor correction capacitors.

- Established in 1978 • Headquarters: Trivandrum, Kerala
- Manufacturing facility: Kannur, Kerala
- Components manufactured: Aluminium electrolytic capacitors, crystals and resistors
- Sectors it caters to: Consumer, power and industrial electronics

Contact details: *Ph: 0091-497-2781922/2780735/781454, gm-kelmktg@gmail.com, www.keltroncomp.org*

6 Victor Component Systems Pvt Ltd

Annual revenue (2012-13): Rs 550.74 million

Pawan Sharma, MD

Victor Component Systems manufactures wound components in its three state-of-the-art manufacturing units spread over an area of 32,000 sq m. Backed by a team of qualified experts, the company caters to the stringent demands of the industry.

- Established in 1989
- Headquarters in New Delhi
- Manufacturing facilities in Noida and Okhla
- Components manufactured: SMPS transformers, ferrites, filters, coils, power supplies, etc
- Sectors it caters to: Consumer electronics, telecommunications, computers, electronics, energy meters, lighting, etc

Contact details: *Ph: + 91-11-26372403, 2637-2404; pawan@victorcomponent.co.in; www.victorcomponent.co.in*

7 Cosmo Ferrites Ltd

Annual revenue (2012-13): Rs 494.10 million

Ambrish Jaipuria, executive director and CEO

Cosmo Ferrites manufactures soft ferrites at its state-of-the-art manufacturing facility with an annual capacity of 4350 tons of ferrite powder and 3600 tons of ferrite products. Since inception, Cosmo has maintained its quality standards and has upgraded its production capacity from 500 MT to 3600 MT per annum over time.

- Established in 1986
- Headquarters in New Delhi
- Manufacturing facility: Jabli, Himachal Pradesh
- Components manufactured: Wide range of soft ferrite cores
- Sectors it caters to: Lighting, solar, power electronics, etc.

Contact details: *Ph: +91-11-49398803, sales@cosmoferrites. com, www.cosmoferrites.com*

8 Desai Electronics Pvt Ltd

Annual revenue (2012-13): Rs 493.67 million

Vikram M Desai, MD

Desai Electronics, popularly known as DEC, manufactures plastic film capacitors at its state-of-the-art plant. DEC recently inaugurated a new manufacturing facility spread over an area of 3000 sq m and the annual capacity of the company now stands at 400 million pieces per annum.

- Established in 1981
- Headquarters and manufacturing facility in Khed Shivapur, Pune
- Components manufactured: Wide range of capacitors
- Sectors it caters to: Lighting (CFL and ballast), power electronics, etc.

Contact details: *Ph: +91-20-24384254/7, mktg@deccapacitors. net, www.deccapacitors.net*

9 BG LI-IN Electricals Ltd

Annual revenue (2012-13): Rs 479.03 million

Rishi Kumar Bagla, director

BG LI-IN is a joint venture between the Bagla Group and LI-IN Electricals of Taiwan. It manufactures various types of flashers and relays. It has a state-of-the-art automatic manufacturing facility built over an area of 4000 sq m. In 2013, the company got recognition for its in-house R&D set-up from the Ministry of Science and Technology, Government of India.

- Established in 2000
- Headquarters and manufacturing facility in Aurangabad
- Components manufactured: Auto and starter relays and flashers, and car security products
- Sectors it caters to: Automotive and power industry

Contact details: *Ph: +91-240-3250408-12, bgliin@baglagroup. com, www.bgliin.com*

10 Genius Electrical & Electronics Pvt Ltd

Annual revenue (2012-13): Rs 399.76 million

Satish Jain, MD

Genius Electrical & Electronics designs and manufactures a wide range of transformers. The company operates from two setups built on an area of about 3900 sq m, equipped with the best German machines. It has an in-house R&D and fully equipped testing lab.

- Established in 1991
- Headquarters in New Delhi
- Manufacturing facility in New Delhi
- Components manufactured: Transformers, SMPS, AC/DC adaptors, etc.
- Sectors it caters to: Power electronics

Contact details: *Ph: +91-11-8113545, 28117826, 28116013, genius66@vsnl.net, genius66@airtelbroadband.in, www.geniusindia.com*

11 General Industrial Controls Pvt Ltd

Annual revenue (2012-13): Rs 389.79 million

Ajay S Chordiya and Ojas S Chordiya, joint MDs

General Industrial Controls Pvt Ltd (GIC) manufactures components like relays, switches, PLCs, etc. It also manufactures various injection moulded plastic and sheet metal components. An ISO 9001:2008 and TS 16949 certified organisation, it has a state-of-the-art plant with facilities for everything from 'design to delivery' integrated under one roof.

- Established in 1972
- Headquarters: Mumbai
- Components manufactured: Relays, switches, PLCs, injection moulded plastic and sheet metal components
- Sectors it caters to: Consumer, power and industrial electronics

Contact details: *Ph: +91-20-30680003/30680004, marketing@ gicindia.com, www.gicindia.com*

12 Incap Ltd

Annual revenue (2012-13): Rs 353.36 million

C Bhagavantha Rao, MD

Incap manufactures electrolytic capacitors and composite polymer insulators. Its state-of-the-art plant is equipped with machinery imported from Japan and Taiwan. Incap has a comprehensive in-house testing facility, with a fully equipped laboratory with T&M instruments of international standards. Currently, the company is executing an order for 400 KV insulators for the Teesta Valley Power Transmission Ltd.

- Established in 1992
- Headquarters and manufacturing facility in Vijayawada
- Components manufactured: Electrolytic capacitors, insulators etc
- Sectors it caters to: Industrial electronics

Contact details: *Ph: +91-866-2842479, incapsales@incaplimited. com, www.incaplimited.com*

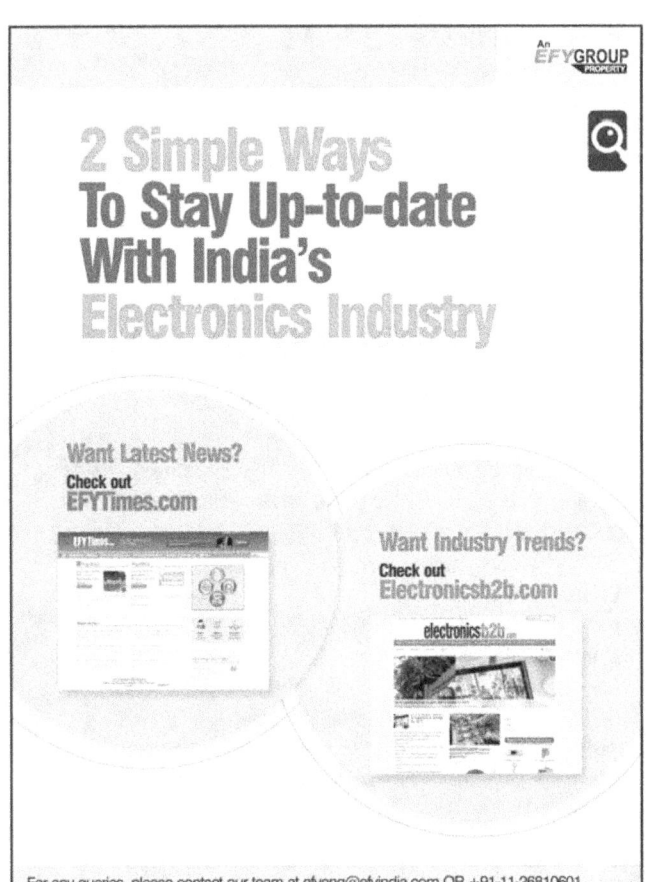

13 Alcon Electronics Pvt Ltd

Annual revenue (2012-13): Rs 323.23 million

Anup R Sachdev, CMD

Alcon is a manufacturer of a wide range of high CV screw terminal type aluminium electrolytic capacitors and film capacitors. Alcon's USP has been its custom design capabilities, and responsiveness to customer needs. Alcon is ISO 9001 and ISO 14001 certified.

- Established in 1977
- Headquarters and manufacturing facility in Nashik
- Components manufactured: High CV screw terminal type aluminium electrolytic capacitors, IGBT, etc
- Sectors it caters to: Power electronics, green energy products, welding equipment, defence equipment and traction and telecom equipment

Contact details: *Ph: + 91-253-2350533; mail@alconelectronics.com; www.alconelectronics.com*

14 Watts Electronics Pvt Ltd

Annual revenue (2012-13): Rs 292.67 million

S Ramakrishnan, director, marketing and sales

Watts manufactures a wide range of resistors, and specialises in a full range of metal film, carbon film, metal oxide, wirewound resistors, etc. The company has five production lines with a total capacity of one billion resistors per year. It entered the lighting industry with CFL, ballasts and other lighting products under the Watts brand name in 2011.

- Established in 1980
- Headquarters and manufacturing facility in Cochin
- Components manufactured: Resistors, capacitors, etc
- Sectors it caters to: Automotive, lighting, instrumentation, electronic appliances and consumer electronics, etc.

Contact details: *Ph: +91-484-2368022, 2373892, watts@wattselectronics.co.in; sales@wattselectronics.co.in, www.wattselectronics.co.in*

15 Solid State Systems Pvt Ltd

Annual revenue (2012-13): Rs 258.76 million

Naushaad Hasan, MD

Solid State Systems, also known as Syscap, manufactures metallised polypropylene film capacitors. During the last 30 years, the company has expanded considerably in terms of volume and product range, manufacturing to international quality standards. It has a fully equipped and automated plant.

- Established in 1972
- Headquarters and manufacturing facility in Bengaluru
- Components manufactured: Wide range of capacitors
- Sectors it caters to: Power electronics, railways and lighting, etc

Contact details: *Ph: +91-80-27971145, ssspl@vsnl.com, www.syscapindia.com*

16 Prismatic Engineering Pvt Ltd

Annual revenue (2012-13): Rs 137.80 million

Rajiv Pandit, MD

Prismatic manufactures ferrite core-based switching power transformers, as well as common mode and differential mode input filter chokes. Prismatic moved into a 929.03 sq m facility in Bengaluru in 2007 and had set up its trading division in early 2008.

- Established in 1994
- Headquarters in Bengaluru
- Manufacturing facilities in Parwanoo and Bengaluru
- Components manufactured: Switching power transformers and filter chokes
- Sectors it caters to: Power electronics

Contact details: *Ph: +91-80-27826274-76, sales@prismatic.co.in, www.prismatic.co.in*

17 Speedofer Components Pvt Ltd

Annual revenue (2012-13): Rs 130.00 million

Randhawa B S, MD

Speedofer Components manufactures soft ferrite cores used in making transformers, inductors, chokes, etc. The company generates about 15 per cent of total revenues from exports. The current production capacity is 100 tons per month. Speedofer is also gearing up for an R&D establishment.

- Headquarters in Noida
- Manufacturing facility in Greater Noida
- Components manufactured: Wide range of soft ferrite cores
- Sectors it caters to: Consumer electronics, LED Lighting, solar, electrical industries, telecommunications, etc

Contact details: *Ph: +91-120-4278911-16; info@speedofer.com; www.speedofer.com*

18 Thakor Electronics Ltd

Annual revenue (2012-13): Rs 114.61 million

Dilip Shah, CMD

Thakor Electronics was founded after Thermax Electronics Ltd was acquired in 2001. Under the brand name Thakor, the company started production of passive components in 2002 at a facility spread over 4468 sq m.

- Established in 2001
- Headquarters in Bhiwandi
- Manufacturing facility in Pune
- Components manufactured: Resistors, CFR, MFR, MOR, etc
- Sectors it caters to: Industrial electronics

Contact details: *Ph: 02522-280277, dds@thakorelectronics.com; mktg@thakorelectronics.com; www.thakorelectronics.com*

19 Cermet Resistronics Pvt Ltd

Annual revenue (2012-13): Rs 106.47 million

Pradeep Khadilkar, CMD

Cermet is a manufacturer of electronics and electrical resistors. It is an ISO 9001:2008 certified company with an annual growth rate of 40 per cent. Its exports make up 20 per cent of its turnover. Cermet has maintained its high-quality standards and is well known for its custom-built solutions. It is all set to diversify and start production of specialised plastic film or film foil capacitors.

- Established in 1990
- Headquarters and manufacturing facility in Pune
- Components manufactured: Film resistors, wirewound resistors and thick film/high voltage resistors
- Sectors it caters to: Auto electronics, industrial electronics, the telecom industry, defence, etc.

Contact details: *Ph: 09975596459/60, sales@cermet.co.in, www. resistorcermet.com*

20 Gujarat Poly-AVX Electronics Ltd

Annual revenue (2012-13): Rs 94.68 million

Tanil R Kilachand, chairman

Gujarat Poly-AVX Electronics (POLY-AVX) is jointly promoted by Polychem Ltd, AVX Corporation of USA and the Gujarat Industrial Investment Corporation Ltd (GIIC). It manufactures capacitors and varistors in its factory, spread across 5000 sq m. During the FY 2012-13, the company recorded a production output of 170.606 million pieces, an increase of 23 per cent.

- Established in 1989 and production commenced in 1993
- Headquarters in Mumbai
- Manufacturing facility in Gandhinagar
- Components manufactured: Multilayer ceramic capacitors in chip and leaded (radial and axial) configurations, single layer ceramic capacitors and metal oxide varistors
- Sectors it caters to: Telecommunications, industrial, medical and consumer electronics applications

Contact details: *Ph: +91-022-22820048, www.polyavx.com*

Leading Electronic Components Manufacturers in India

Radius Industries

Inderpreet Singh Bindra, director

Radius Industries is an ISO 9001-2008 certified company, and offers international quality reliable electronic components under the brand name Radcom. At Radius, the emphasis is on quality and service. It has a full-fledged sales and marketing team with a strong network of competent distributors and dealers spread across the country.

- Company was established in India in: 1998
- Headquarters: New Delhi
- Components manufactured in India: MOSFETs, MOVs, NTCs and ceramic chip capacitors, LEDs, resistors, rectifiers, bridges, Schottky diodes, rectifiers, etc
- Brands it offers: Radcom
- Sectors it caters to: Communications, power electronics, consumer electronics, lighting, etc.

Contact details: *Ph: 011-26386843, 26383057, 9312068847, sales@radcomsemi.com, www.radcomsemi.com*

Rank Infotech

Prakash Bharwani, director

Toyo Connectors and Cables is an ISO-9002: 2008 company providing all kinds of connectivity solutions with a range of over 6000 products. The company manufactures over 60 types of relimates and cable assemblies as per customer specifications. It is known for its quality—its rejection rate is less than 0.5 per cent. Its products have international approvals like UL, CE, VDE and RoHS.

- Headquarters: Mumbai
- Components offered in India: Sensors, D-Sub, VGA, mini and micro USB, HDMI, serial convertors, adaptors, heat shrink sleeves, antennas, etc.
- Brand it offers: Toyo
- Sectors it caters to: Industrial automation, chemical, electrical, electronics, communications, avionics, automotive, aerospace, medical, etc.

Contact details: *Ph: 022-23882324/25 (Mumbai), 080-40987129/9740508439 (Bengaluru), toyoconnectors@gmail.com, www.toyoconnectors.com*

Leading Electronic Components Distributors in India

In alphabetical order

3B Semiconductors Pvt Ltd

Sachin Bhalerao, director

3B Semiconductors Pvt Ltd is a premier electronic components distributor that provides complete support to customers during the life cycle of the product.

With wide experience in the electronics field, the company offers the best possible technology at an affordable price. Its strong technical network helps customers to select the right components for their needs. Through its 3B Semi Design system, the company focuses on providing designs to those who can sell them and on developing designs that are sell-able.

- Headquarters in: Mumbai
- Components offered in India: Diodes, transistors, SCRs, mosfets, IGBTs, LDOs, MCUs, FPGAs, memory, LEDs, etc.
- Brands it offers: ST, CREE, Glacial, NXP, ON, TI, Infinieon, FSC, Hermai, Atmel, Cypress, Micron, Microchip, etc.
- Sectors it caters to: Industrial and consumer electronics

Contact details: +91-22-24039287/88/89, info@3bsemi.com, www.3bsemi.com

Alfa Electronic Components

Ashish Mehta, CEO

Alfa Electronic Components has been marketing the most comprehensive and wide range of LEDs and LED displays over the past 20 years. Instead of limiting itself to being just a distributor or sales outlet, Alfa has stretched itself to grow in all directions. It is an authorised distributor for Twilight Opto Electronic Co Ltd, Taiwan. It offers prompt technical support, comprehensive datasheets and helps customers choose the right product for their applications. Value addition plays a pivotal role in its operations and is driving its business forward, creating a distinct position for Alfa in this competitive market.

- Company was established in India in: 1993
- Headquarters: Mumbai
- Components offered in India: LEDs and LED displays
- Brand it offers: Twilight
- Sectors it caters to: Process control instruments

Contact details: Ph: 022-26840075, alfaopto@gmail.com, www.alfaopto.com

CEE PEE Electronics

Paresh Premji Gala, director

Viral Gala, director

Situated in the heart of south Mumbai, CEE PEE Electronics deals in all types of industrial components and is able to provide complete packages to its customers. It also stocks robotic components and parts. Being situated at a strategic location, the company is able to source components from across the globe. It believes in value-based pricing for its large customer base. CEE PEE is committed to continuous improvement in quality management to ensure customer satisfaction.

- Company was established in India in: 1993
- Headquarters: Mumbai
- Components offered in India: DIP and SMD, IC transistors, diodes, LEDs, connectors, robotics components, AC DC fans, displays, etc
- Brands it offers: Texas, ST, Microchip, Raxon, JHD, Maruwa, HYDZ and HXD
- Sectors it caters to: Government, PSU, medical electronics, etc

Contact details: Ph: +91-22-23825859/66347509/ /23890521, 9870042030 ceepee@bom3.vsnl.net.in, cp_ceepee@yahoo.co.in

Cirkit Electro Components Pvt Ltd

Gulabchand Hariya, director

Cirkit Electro Components Pvt Ltd is among the leading independent stockists and distributors in India, with 25 years' experience. It is a reputed supplier in the electronics industry with proven credentials both in the domestic and overseas markets. The company specialises in supplying all types of components in the kit form to OEMs and EMS companies. It has wide sourcing contacts with major distributors and manufacturers from America, Europe, Japan, Singapore, Hong Kong and China.

- Company was established in India in: 1990
- Headquarters: Mumbai
- Components traded in India: ICs, semiconductors, MOSFETs, IGBTs, triacs, SCRs, transistors, capacitors, diodes, LEDs, LDRs, LCDs, tantalums, sensors, etc
- Brands it offers: IR, Vishay, Fairchild, STMicro Electronics, Microchip, Atmel, PI, ADI, Inchange, SM Micro, CDIL, Cheng Capacitor, etc
- Sectors it caters to: Industrial and consumer electronics, power, lighting, etc.

Contact details: Ph: 022-23877777/23800888, 9322234370/ 9322791112, cirkitelectro@gmail.com, www.cirkitelectro.com

An EFY GROUP EVENT

www.efyexpo.com

BOOKINGS OPEN

FOR

2nd Edition

Electronics ForYou expo 2014
WESTERN INDIA
Nov 26-28, 2014
Mumbai

ONLY ELECTRONICS SHOW CATERING TO WESTERN INDIA

We Promised 1500 But Delivered 3000 Business Visitors*

(We love it when we under-promise and over-deliver)

TESTIMONIALS OF EXHIBITORS

The response was exceptionally good. We were pleasantly surprised with the turnout of quality visitors.	The turnout of people has been exceptional. We have got quality leads.	There were customers looking for the right solutions for highly technical products in the areas of space, military applications, etc.	We have gathered some good enquiries.
Sujata Soparkar, **MD**, INTRON	Dinesh Patkar, **country manager**, Quectel	Nariman Mehta, **CEO**, Progressive Engineers	N Y Patil, **MD**, Kyoritsu Electric India

(* Actually, more than 5,000 visited, but from these only 3,000-plus were Business Visitors)

BOOK NOW TO AVAIL EARLY BIRD DISCOUNTS AND A GOOD LOCATION FOR YOUR BOOTH!

Special 'Combo Packs' are available for booking booths at EFY Expo West (Mumbai), EFY Expo India (Delhi), IPCA-EFY Expo (Pune) and eRocks (Bengaluru). Call Arun at +91 88000 94213 or email us on efyexpo@efyindia.com

Leading Players

Componix India

Chetan Ajmera,
director

Componix is among the fastest growing organisations in the distribution of electronic components. The company focuses on distributing reputed brands and has business agreements with a few leading suppliers from Taiwan, Korea, China, USA, etc. Its emphasis is on expanding its product portfolio to serve existing customers. Componix supports a wide range of electronic components and supplies them at competitive prices. It also offers consolidation services from diverse components manufacturers along with MODVAT/zero excise benefits.

- Company was established in India in: 2004
- Headquarters: Mumbai
- Components offered in India: Sensors, rectifiers, capacitors, triacs, diodes, etc.
- Brands it offers: Samwha, TSC, Sisemi, Jifu, Weidy, Haudy, etc.
- Sectors it caters to: Lighting, power, auto electronics, telecom, etc.

Contact details: *Ph: +91-22-23827771/23827772, chetan@componixindia.com, www.componixindia.com*

Millennium Semiconductors

Haresh Abichandani,
MD

Millennium Semiconductors is one of the fastest growing distributors of electronic components in India. It is committed to offering the broadest selection of in-stock electronic components, as well as providing the best service possible to its customers. The company has its own R&D centre supported by a team of dedicated R&D professionals to create innovative solutions. It has branch offices at Delhi, Bengaluru, Hyderabad, Ahmedabad, Chennai, Coimbatore and Mumbai, in India, with overseas offices in Singapore and Shenzhen (China).

- Company was established in India in: 2004
- Headquarters: Pune
- Components offered in India: Sensors, relays, SCRs, circuit protection devices, microcontrollers, regulators, etc
- Brands it offers: Littelfuse, Melexis, PIC, Hongfa, Digi, Nuvoton, Spansion, National Chip, Quectel, AVX, Fairchild, etc
- Sectors it caters to: Automotive, industrial, consumer electronics, defence, etc

Contact details: *Ph: +91-20-27484800, 27484900/274845000, info@millenniumsemi.com, www.millenniumsemi.com*

Rabyte Electronics Pvt Ltd

Rajiv Batra,
managing
director

Rabyte is among the leading distributors of electronic components in India with ISO 9001:2008 certification. It offers components with advanced technology from various reputed manufacturers across the globe. It acts as a single window source to all its customers for their electronics solutions requirements, including design services, box-build solutions, supply chain management, etc. The company has expertise in supporting customers with kitting services for product BOMs at cost-effective prices.

- Company was established in India in: 1986
- Headquarters: Noida
- Components offered in India: Semiconductors, passive components, interconnect and electro-mechanical components, displays and modules
- Brands it offers: Renesas, Hynix, Infineon, NXP, Sony, Rohm, Littelfuse, Liteon, Osram, etc.
- Sectors it caters to: Automotive, consumer and industrial electronics, lighting, IT, telecom, etc.

Contact details: *Ph: +91-120-474-6000, enquiry@rabyte.com, www.Rabyte.com*

Swingtel Communications Pvt Ltd

Moiz Manasawala,
MD

Swingtel Communications, which is part of an international group, is among India's leading distributors with a global sourcing network. Swingtel designs and develops solutions and offers the best-in-class components—both active as well as passive—to Indian customers. The company has been continuously innovating and evolving to take on the challenges of the dynamic business environment and add value to its end customers.

- Company was established in India in: 2004
- Headquarters: Mumbai
- Components offered in India: Microcontrollers, ICs, MOSFETs, triacs, optocouplers, regulators, transistors, LEDs, etc
- Brands it offers: ST, NXP, Texas, Goodark, HEL, KEC, Atmel, etc.
- Sectors it caters to: Automotive, lighting, power, instrumentation, telecom, etc

Contact details: *Ph: +91-22-2200 0123, info@swingtel.com, www.swingtel.in, www.swingtel.com*

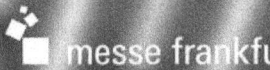

66Govt is doing enough to support domestic manufacturing; it is now time for the industry to take action

In January 2014, Trust Research Advisory named South Korean electronics major Samsung as India's most trusted brand. When a brand gains the trust of consumers, it also gains market share and product acceptance. Samsung has been one of the first companies to invest under the government's initiative to boost electronics manufacturing in India.

Srabani Sen of *Electronics Bazaar* caught up with Ravinder Zutshi, deputy managing director, Samsung India Electronics, to understand how Samsung is contributing to the electronics manufacturing ecosystem in India.

Ravinder Zutshi, deputy managing director, Samsung India Electronics

EB: What has Samsung achieved under your guidance?

I have been part of Samsung Electronics' journey to success in India from the very beginning. One of my early efforts was to set up a strong distribution system in the consumer durables space, which has been the foundation of the company. Over the last several years in Samsung, I think my biggest contribution has been to make Samsung one of the most aspirational brands in India and to establish it as a 'top-of-mind consumer product' company, across product categories. We have been successful in creating a robust ecosystem consisting of partners and consumers and we have grown severalfold during my tenure.

EB: How is Samsung contributing to the electronics manufacturing ecosystem in India?

Our manufacturing is based on localisation. A large number of local vendors work with us in Noida and Chennai, where we have two large manufacturing units for consumer electronics and mobile phones. While the Chennai facility manufactures only consumer electronics products, at Noida, we manufacture both consumer electronics and mobile phones. Wherever possible, we procure electronics components and other raw materials for our products from local vendors situated around our facilities. However, due to the high end nature of our business, a large part of the components are still imported.

EB: Any plans of Samsung participating in semiconductor manufacturing in India?

No, we have not been involved in any such discussions.

EB: What is your take on the two upcoming fabs in India?

Well, that is a good beginning. I think the government has done a great job in terms of inviting corporate houses to invest in fabs. I am sure this will go a long way in developing a very robust ecosystem for components that are not available in India at the moment. We are looking forward to such facilities in India with the hope that they will help in reducing the import burden of some components.

EB: What is Samsung's investment under the Modified Special Incentive Package Scheme (MSIPS)?

Under MSIPS, Samsung has invested Rs. 8000 million to enhance the production line for its Galaxy series of smartphones and tablets and, eventually, to increase productivity at the Noida plant. We made the investment last year in February, and the work related to this investment finished by October.

EB: Are the mobile phones manufactured in Noida sold in the domestic market?

By and large, India-manufactured mobile phones are sold in the domestic market. However, some are also exported. This is a big initiative by Samsung as local manufacturing helps to create and expand the market in India and helps us to cater to the strong domestic demand. We want to showcase the manufacturing capability we have in India.

EB: The government is currently supporting electronics manufacturing. Do you think it is on the right track?

The government has initiated several measures to improve the industrial climate and manufacturing ecosystem within the country. This will certainly help in building the confidence of the industry and Samsung fully supports these initiatives. The most important initiative taken by the government in this regard is the announcement of the National Manufacturing Policy (NMP) under which the implementation of MSIPS was a very bold step.

The action plan of the government seems pretty clear—revive the investment cycle, sustain reforms, create skilled jobs and increase exports. Earlier, it did seem that India had missed the bus but now it appears to be on the right track. The government has an ambitious plan up its sleeve, but will need to be really pushed forward.

EB: But despite government support, we still do not see much of manufacturing happening in India...

The government reacted rather late, but whatever it has done in the last two years is commendable. I am sure more and more companies will come forward with investment proposals. I believe there are lots of applications pending with the Department of Electronics and IT (DeitY).

EB: Domestic companies complain that components that are being imported are cheaper than what they are producing in India...

Cost is not the issue. The issue is that there has to be demand in order to build up that capacity. And capacity build-up will only happen when people start investing in local production. So, once you start having local production, the components industry will surely thrive. Anybody who puts up a components factory has to look at scale, which is very important. And scale will only

come when local manufacturing happens. So you see it is an interlinked story. But I am sure that with these new government initiatives, people will definitely invest in local manufacturing of electronics hardware.

The central government and the state governments are doing enough to support domestic manufacturing; it is now time for the industry to take action. So, everybody needs to think and start investing in India.

EB: What is your current strategy in India?

Our strategy is to grow our business significantly across categories by introducing several innovative products with differentiated features in India, including for the entry-level segment. We provide the most innovative products with the best possible technology to our consumers.

At Samsung, our constant endeavor is to meet the requirements of the consumers. We understand the importance of timely and relevant communication, and we use all the available media in a creative and innovative manner—be it television commercials, digital and social media, or interacting with our consumers extensively through our vast dealer and distributor network.

> " Our strategy is to grow our business significantly across categories by introducing several innovative and customised products in India, including for the entry-level segment. We provide the most innovative products with the best possible technology to our consumers.

EB: What is your focus this year?

Our focus this year is to strengthen our existing businesses by adding new, innovative and technologically advanced products, yet at an affordable price. We will also focus on developing a bigger consumer base for our products. Besides, our aim is to attain the leadership position in consumer electronics and smartphones. We will also focus on expanding our distribution network and create more exclusive stores throughout the country to provide our consumers with better product experience.

Innovation and technology have always been our differentiators, and we will work on these strengths much more as we move forward with our business plans. **EB**

Dixon Technologies aims to be one among the top five global EMS companies

In the next two years, Dixon plans to explore the export market. It also plans to increase its capacity to penetrate unexplored segments of the domestic market

By Srabani Sen

Over the last two decades, Dixon Technologies India Pvt Ltd has evolved into more than just an electronics manufacturing services (EMS) company. Headquartered in Noida, Dixon is the largest Indian EMS provider, focused on delivering high quality, cost effective solutions in the domestic consumer electronics, lighting, set top boxes (STBs) and home appliances space. It is one of the few Indian EMS players with full scale capabilities in product design, global procurement, turnkey manufacturing, logistics and reverse logistics and strong back-end support.

"We look at ourselves as a complete solutions provider, as we offer services that begin right from designing the product to assembling it. We even handle the logistics and reverse logistics, wherein we take care of after-sales service and the refurbishment of the products," says Sunil Vachani, chairman and managing director, Dixon Technologies.

Established in 1993, Dixon has grown rapidly in the last two decades, and is recognised for its low cost manufacturing capabilities, proven design expertise and established customer base.

Maintaining a competitive edge

The manufacturing process at Dixon is vertically integrated—it has its own plastic injection moulding plant, sheet metal plant, and even manufactures its own wound components. It will soon start manufacturing some other components as well. "There is

Sunil Vachani, chairman and managing director, Dixon Technologies

Atul Lall, co-founder, CEO and deputy managing director, Dixon Technologies

no Indian or multinational company in this space today that offers these kinds of packages or solutions to the customers," says Sunil.

"Our vertical integration and scale of production make us one of the few low cost producers, allowing us to pass on the benefits of low cost production to our customers," says Sunil. "This, coupled with our focus on processes, quality and continuous improvement in systems, helps us to maintain a competitive edge," he adds.

The journey so far

With a meagre investment of just Rs 2.5 million, Sunil started Dixon in 1993. "Thanks to our customers and the great team that the company has, we crossed a turnover of Rs 10 billion last year -- within a short span of 20 years," says Sunil.

Recalling the initial hurdles faced, Sunil says, "The biggest challenge was to get funds, as back then, convincing the bankers about the viability of the business was a difficult task." EMS was a relatively new concept in India in the nineties, and it was a risk that Sunil was undertaking. As expected, the company did not make a profit in the first year.

Talking about other challenges faced in the initial years, Sunil says, "Getting the right people was a major challenge. This is also an important aspect of a company's success, as it is the people who will implement the right processes. But I was lucky to have Atul Lall as my partner and co-founder, who is currently the CEO and deputy managing director of the company. He has been a pillar of strength for the company, and is responsible for creating systems, processes and bringing the right people on board," informs Sunil.

Two decades back, Dixon started its operations from a rented unit in Noida. Today, it has six manufacturing plants and a workforce of almost 4000 people. In 1993, the company started assembly operations for consumer electronics products such as CTVs and VCRs. Back then, two of its OEM customers were LG and Philips. Between 2000 and 2008, the company developed capabilities in PCB manufacturing and the final assembly for complex products. It began assembly operations for new products like DVDs, CFL bulbs and

STBs. It also acquired new customers such as Kaon Media, Godrej, Videocon and Onida. Gradually, by 2010, it developed low cost original design manufacturing (ODM) solutions for products like CTVs, DVDs and CFL bulbs.

In 2010, Dixon entered new categories like LCD/LED TVs, energy meters and power inverters. During the same time, it also started manufacturing home appliances like washing machines and induction cooktops, apart from venturing into LED lighting. During this period, Dixon earned the trust and faith of OEMs like Panasonic, Toshiba and Landis Gyr.

"Since our inception, we have been driven by a clear vision, which helped us in carving a distinct space for Dixon at the forefront of the electronics manufacturing and contractual manufacturing industry in India. This clearly defined mission helped Dixon to acquire and retain its EMS clients in India and globally," says Sunil.

Alliances and acquisitions

To offer the optimum cost advantage to its customers, Dixon promoted Bhurji India Pvt Ltd in 2008. It channelised its sheet metal and plastic injection moulding services through this firm, thereby making it a world class facility for sheet metal and engineering components in India.

Dixon had promoted My Box Technologies in 2009 along with

Kharabanda family. While My Box designs and develops STBs, Dixon manufactures them at its facilities. "Dixon and My Box have been complementing each other, and have been instrumental in laying the foundation for a successful partnership with their expertise in manufacturing STBs for both the digital cable and satellite TV markets," says Sunil.

Dixon Appliances Ltd was also established in 2009 with the aim to manufacture home appliances. This company is headed by Mr. Sahil Vachani. True to its commitment, the company today stands proud as one of India's leading home appliance manufacturers with a customer base of leading brands like Godrej, Haier, Electrolux etc.

Strengths and achievements

Talking of the strengths of the company, Sunil says, "Maintaining customer relationships is our major strength. Our foremost focus is on building customer partnerships by providing products and services of the greatest value through innovation and excellence."

This has been achieved by the firm attaining Six Sigma capabilities in all key processes. "We also give emphasis to strengthening supplier partnerships through good communication and recognition, all of which are our strengths. Conducting business with uncompromis-

One of Dixon's manufacturing facilities

Dixon's assembly unit

ing integrity is another of Dixon's strengths," says Sunil.

Dixon is also driven by a passion to innovate and excel, and the constant desire to come up with something better each time.

In fact, the achievement that Dixon is most proud of is its ability to retain customers, all of whom have been with the company since it started. "Not a single customer who has been with us since the company's inception two decades ago has left us," says Sunil. He is also proud of the fact that the company has grown from 40 people to 4000. "We have a team which is very motivated, charged and willing to do something new every day," adds Sunil.

Dixon has diversified its product portfolio to a great extent. Explains Sunil, "Diversification is important for a company to grow faster. Products are very cyclical, and have ups and downs. So if you just depend on one product, there are very high chances of failing as an organisation. I am very satisfied with the fact that we now are a multi-product, multi-location business, which hopefully, will hold the company in good stead in the days to come," adds Sunil.

Another achievement of Dixon's that Sunil likes to share is its capability to design STBs, which many considered an impossible task in

India. "I remember, when we were trying to design STBs five years back, some of my Korean OEMs said that I was wasting money and time on it, because many companies had tried to design an STB but failed. But we believed that Indian designers, along with hardware and software engineers, had the capability to design these boxes, and finally we did. So that is definitely a way forward. We cannot just take designs from China, reverse engineer them and sell them in India," says Sunil.

Areas for improvement

Sunil feels that despite operating for two decades, the company is still learning and trying to improve on certain areas that need attention. Design is one aspect the company needs to strengthen further, so that it can be capable of developing products at frequent intervals since the industry is changing very fast.

Another area in which the company feels it can improve further is with regard to processes. "In this aspect, a company needs to evolve every day because in every activity we do, there is always the scope to do things better," says Sunil.

Sunil also feels that the company needs to improve its infrastructure. "Although we have seven facilities,

there is definitely scope for more automation. Today, if you go to China, you can see how robots work in factories—many processes and activities are automated. So, obviously our infrastructure needs to become world class in the years to come," he says.

Sectors catered to

Dixon is present in four different segments—consumer electronics, lighting, STBs and home appliances—besides the general EMS category that covers many other products. In consumer electronics, its main focus is on televisions, LED TVs, DVD players and home theatre systems. In lighting, Dixon is the largest player in CFL bulbs and circuits, claims the company. It is also in the LED lighting and LED driver circuits space. In home appliances, it claims to be the largest manufacturer of washing machines. And it is India's first company to have started manufacturing STBs based on its own designs.

State-of-the-art infrastructure

Dixon has six facilities—three in Uttarakhand and three in Noida, Uttar Pradesh, spread over 92,903 sq m. "We have 11 SMT lines, and the largest capacity for auto insertion. We also have infrastructure for manual insertion and plastic injection moulding, right from 80 tonne machines to 1600 tonne machines. We have sheet metal infrastructure right up to 160 tonnes and have our own moulds for washing machines," says Sunil.

Dixon has the capacity to manufacture 1.5 million LED televisions per year, about 60 million circuits per year (which is being increased to 80 million circuits), and 40,000 washing machines per month (which is about half a million per year). It also has the capacity to manufacture almost 2 million STBs per year.

Dixon will soon establish one

more unit in Noida, next to its existing plant, as it feels the need to do more backward integration and increase its infrastructure for reverse logistics. "We need this unit in order to increase the production capacity of our existing plant in Noida," says Sunil.

Sustainability strategies

Sunil believes that a company has to build strong sustainable and competitive strategies to survive in this industry. Some such strategies include creating a scale by producing Six Sigma quality, having processes that are difficult to emulate, and offering the lowest cost to customers. "A combination of these sustainable strategies gives us a competitive advantage," says Sunil.

Dixon's vision is executed not only by the management but is shared by everybody working in the company. "Once people believe in that vision and once they see that I am sincere about what I want to accomplish, a lot of things fall into place. And as the MD and chairman of the company, my key task is to convince my people about the vision of the company. In fact, my vision is never expressed in terms of numbers, because if I tell my people that my vision is to grow at a rate that is unrealistic, they will not believe in me. Also by defining your vision in numbers you tend to limit the potential for growth. Instead, my vision is to become the most respected company in the field of electronics manufacturing services. And that can happen by becoming the company that delivers the best quality in the industry—our processes are the best in the industry, our cost structure is the lowest and we are exceeding our customers' expectations in every parameter. So we are making our vision come true each day," says Sunil.

The second most important strategy of the company is to innovate every day. Sunil believes that a company will die if it stops innovating and becomes stagnant. "In our industry, we cannot survive or be in the competitive race if we do not innovate—whether it is with products, processes or systems," he says.

Another strategy that the company follows is to train people to meet the challenges an EMS company faces. "We have our in-house training facility and a soft skills training company that coaches anyone joining the company over a period of 30-45 days. We train them on theory, component insertion, and also give on-the-job training," says Sunil.

The company follows a focused and structured review process, ensuring a well-documented feedback process for all customer complaints that come up. The review system operates at all levels, starting from the workers' level on a daily basis, to the managerial level on a weekly basis, and the board and senior management level on a monthly basis.

Future plans

Dixon's focus areas are lighting, STBs, home appliances and LED televisions. While all these markets are growing, the penetration of these products is still very low. So Sunil feels that, in the near future, the company will have to increase its capacity.

Dixon is the largest Indian EMS company. Sunil would like to turn it into one of the top five global EMS companies, competing with the global giants with its focus on innovation, quality and execution. "I know it's a very grand vision, but only if we dream of it will we be able to achieve it," he says.

In the next two years, Dixon also plans to explore the export market. "I know it is a different ball game, as the international market is very different from the domestic market, and customer expectations in terms of quality and cost will also be very different from what we have been experiencing in India. But as a team we have to gear up to face the challenge," concludes Sunil. ▣

KEY FACTS AT A GLANCE	
Year of establishment	December 1993
Turnover (2012-13)	Rs 10 billion
Workforce	Around 4000
Manufacturing units/ plant locations	Six facilities in Uttarakhand and Uttar Pradesh, spread over 3716 sq m, 11,148 sq m, 7246 sq m, 21,368 sq m, etc
Major customers	LG, Toshiba, Panasonic, Philips, Godrej, Landis Gyr, Videocon, Havells, Dish TV, Walmart, Akai, etc
Product range	Colour TVs, DVD players, LCD solutions, washing machines/washers, juicers/mixers/grinders, multimedia speakers, STBs, CFL lights, Motion Picture Editors' Guild cards, LED lighting solutions, etc
Sectors catered to	Consumer electronics, lighting, STBs and home appliances
Services offered	Design and assembly of products, backward integration for a lot of products that require plastic injection moulding and sheet metal, logistics and reverse logistics.
Contact details	B-14 and 15, Phase II, Noida, Uttar Pradesh 210305, Ph: 91-120-2562639, 2562820

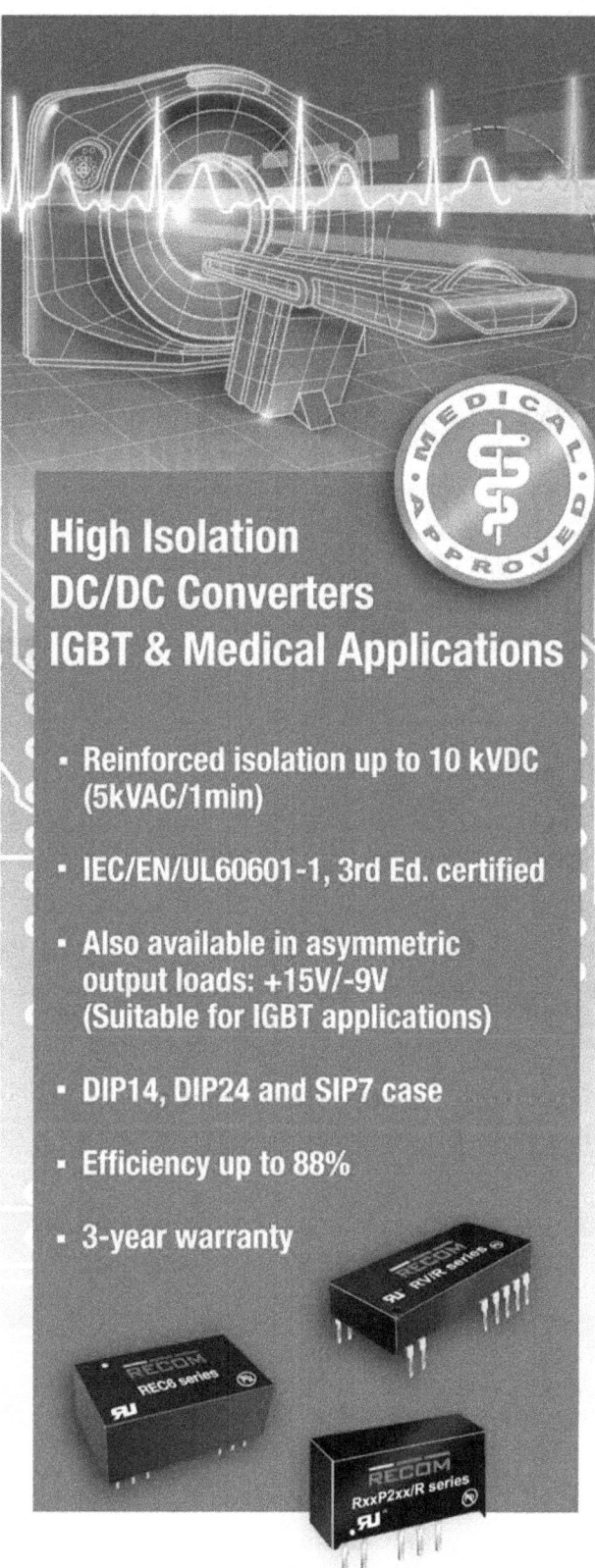

"Happiness on the faces of everyone around me is my goal

He is polite, yet straightforward and a strong willed person. Meet the soft-spoken **Ajay Goel, CEO, Goldwyn Limited,** who took over the helm of the company at the young age of 23. In a tete-a-tete with **Kartiki Negi** of *Electronics Bazaar,* he retraces his 22-year journey with Goldwyn and talks of how, despite several deterrents, he chose to stick with the LED business

I was born in 1969 into a middle class family that had strong values. I was brought up in Delhi, where my father worked with the Ministry of Defence, Government of India, while my mother was a homemaker. My childhood, spent with my elder sister and younger brother, was very sheltered. As a child, I had a keen interest in cricket and was even part of the school and college cricket teams.

I am a true Delhiite. I completed my schooling from Harcourt Butler School, located in the heart of the city, and by God's grace, I was always among the top rankers in the class. I went on to do my bachelor's degree in electronics and telecommunications engineering from Aligarh Muslim University. After completing engineering, I wanted to start my own business but life had some other plans for me.

Professional journey

After finishing my engineering, I joined Automotive Alliance in 1991 as an engineer trainee. Within a year, I was shifted to Goldwyn in 1992, which was a great opportunity given to me by the board of directors. Though I am an employee of Goldwyn, I have always treated the company as my own.

When we had started Goldwyn, it was a very small organisation with only six employees. I have been associated with the company for about 23 years now and have seen several ups and downs, but I am proud that the company has emerged as a winner. I have seen it grow in terms of both revenue and people. Over several stages, we grew from six people to 50, then to 100 and now, we are over 150 people.

Initially, we were engaged in the automotive parts business but eight years back, all of a sudden, one of our customers in the UK, to whom we were supplying parts for the UK metro, asked us to also supply LED parts. That is how we ventured into the LED business.

However, this journey into the field of LED lights was not a cakewalk, as it was a new and evolving field. The products that we make today might not be what we will be making two years or even one year from now. When we started out, LED lights were extremely expensive. Back then, convincing Indian and foreign customers to opt for these lights was not easy.

Today, our major customers hail from Germany and other European countries, where quality plays a very important role. To have supplied to these markets for more than seven-eight years shows our credibility, in terms of our quality and commitment. Initially, though defects were reported in a few products, we were able to rectify them and get those orders back from the same customers.

At this point, I would like to mention two major international projects that I found really satisfying. One was at the Chinook helicopter base in UK, which we undertook about six years back, and it was quite commendable that we could reach out to a customer of that stature. The second one is at the port of Houston where our lights are working at a height of 61 metres above the ground.

We have also done lots of projects across India. I am very proud of the fact that our products are being used at the border by the Indian Army to stop infiltration. It is extremely satisfying for me that we are able to contribute towards protecting our nation, in whatever way we can.

Challenges faced

The motivation to remain in the LED business was initially very low because of the slow growth rate. However, when I realised that Philips' CEO was able to really fast-track LED growth across the entire world, including India, I felt that Goldwyn had a role to play. Initially, we were not able to achieve the volumes we were aiming at but, gradually, the cost of LED lights has been dropping. Also, to a certain extent, the awareness is also increasing among customers—that even if the initial investment they are making is high, the payback period is short. Now, we are getting new customers almost every day.

The R&D team is the backbone of any business. Initially, we did not have one, but in a landmark move about 10 years back, we set up our own design team. However, developing the R&D team was a major challenge, as skilled hands are still scarce in the LED field. Since then, there has been no looking back. Five years back we also entered the field of electronics design. Today, we do the entire mechanical as well as electronics design of our products. About three years back, we went into the optical side of the business as well. At present, the strength of the company lies in its designing and manufacturing capabilities. All the design techniques have been developed in-house. Today, when I see what our full-fledged R&D team has achieved, I feel very proud and satisfied.

Setting up the manufacturing plant in Noida was quite a task. The entire process—from getting the land to setting up the production floor with modern machines and getting the right people to run them—was fraught with challenges and hardship. But with the help of my team, we could resolve every problem that came our way. Today, Goldwyn's plant is spread over 9000 sq m in Noida, and manufactures all types of LED luminaires. Our manufacturing philosophy is to produce all critical components and assemblies in the plant, so as to maintain high quality standards and guarantee the reliability in our LED lighting products. We will also invest in manufacturing metal casings with high speed turret punch press and power coating plant. This move will help us to ensure excellent quality products in shortest possible time.

I hope I will be able to continue fulfilling the aspirations of the board of directors of the company, by taking up new challenges and overcoming them successfully.

Ajay Goel with his colleagues

Management style

I believe in delegating responsibility in a well-documented and structured manner. My management philosophy is to give full authority to people to work in their own way.

Motivation plays a vital role in how well people work. If we are able to elicit new ideas from members of the team and implement them, then that it is a good motivating factor. You will be surprised to know that some of the production techniques we follow have been devised by an employee who is not even an engineer. I strongly believe that the person who is dealing with a particular task every day can give us good suggestions, and I make sure that those who make valuable suggestions are well taken care of.

Besides, LED is a new technology, which in itself is a motivational factor for everybody, whether they are in the design, manufacturing or marketing teams.

Turning points and achievements

Though life has been very kind to me, there have been some incidents that have changed the course of my life. I was quite young, just 17, when I lost my mother. Though tragic, her death impacted my life in a positive manner. Earlier, I was not focused about what I wanted to do in life but after she passed away, I started taking life seriously.

Then in 1992, within a year of my joining the organisation, I was given the flexibility to run Goldwyn—it was a small firm then. That was a high point in my career that I can never forget. I was given complete responsibility for the organisation by the board of directors.

As the organisation grew, I too grew both personally and professionally. I have been associated with the special economic zone (SEZ) movement in the country and, in 2002, I was selected to join the panel of entrepreneurs by the Government of India. We advise the government on how to improve the SEZ policy and SEZ Act. This work really gives me satisfaction, as I not only come to know about the challenges being faced by the industry but am also actively involved in finding out ways to overcome them.

Ajay Goel with his family during a vacation

Success, a subjective term

Success is a very subjective term for me. Whatever I term as 'successful' today might not hold any relevance tomorrow. What is considered 'successful' depends on an individual's psychological state, at that time. In my view, those who feel that they are successful have perhaps stopped growing, personally. So, I don't give this term much importance.

Rather than being remembered as a successful man, I would like to be remembered as a person who tries to help everybody. Happiness on the faces of everyone around me is my goal.

I have always advised young and budding entrepreneurs to make their own plans, not in their minds but on a piece of paper. Being ready with the complete roadmap of your plan, and going out to achieve what you set out to do, is what I feel people should do to achieve success.

WHAT I WOULD LIKE TO CHANGE...

- *In the world:* Reduce the disparity between countries
- *In the country:* Reduce the disparity between the rich and the poor
- *At my workplace:* Have a less hectic work schedule
- *In myself:* I want to be full of energy when I start my day

THESE ARE A FEW OF MY FAVOURITE THINGS...

- *Music:* Old songs
- *Food:* Homemade
- *Film:* Old Bollywood films
- *Book:* Mahatma Gandhi's autobiography
- *Hobby:* Cricket
- *Holiday destination:* Kashmir
- *Actor:* Dev Anand
- *Actress:* Hema Malini
- *Role model:* Mahatma Gandhi

My family is my strength

My family is my biggest strength. I attribute my success to my father, who has worked hard over the years to make me what I am today. He is a passionate worker and very honest about his work. I think I have acquired that quality from him.

Other pillars of my strength are my wife and children. They understand that my profession requires more hours of work than is normal. I am very thankful for their patience and understanding when I may not be able to come home every day at 7 pm in the evening. They have supported me and stood by me through all the ups and downs. When you have a supportive and understanding family, you tend to become more responsible towards it and make an effort to take care of all the family members' desires and happiness. A supportive family also helps you to concentrate on the professional front better.

I have been married for almost 21 years now. My wife has always supported me emotionally, whenever I have had low periods in my profession. She is a commerce graduate, is ICWA certified and is engaged in teaching. She is also actively involved in a lot of social work.

I have two children—my elder daughter is studying to be a chartered accountant, while my son is studying automotive engineering from Manipal Institute of Technology. He is also passionate about engineering but in a slightly different field than mine. He, along with a team, is building a car called Formula Manipal. It will be showcased in Germany, within two years.

Furure plans

All this while, I have been working mostly in the export arena, whether it was for automobile parts earlier, or LED parts today. We now want to explore the domestic market as well because we feel the LED lights market in India has immense potential. It is also high time that we educate people here to shift to LED lights as it is an energy-saving and cost-saving technology.

I strongly feel that the LED market is poised for growth and that, in the coming two years, more than 60 per cent of the lights in the market will be LED-based. At present, the Indian market has a growth rate of 40 per cent. However, this growth is nothing compared to India's potential. In the coming years, I aim to achieve 50 per cent growth for Goldwyn.

As far as my personal life is concerned, I would like to go slow, post retirement. I want to contribute and work towards the improvement of the industry. In all possible ways I would like to help the government in its endeavour to boost manufacturing in India. If given a chance, I would also like to give suggestions to the government on how to clean up our polluted rivers, or work towards a better taxation policy.

Apart from this, I will more actively follow my hobby of playing cricket. As of now, I get very little time for it but will make sure that I practice on the pitch a little more once I've retired. ▣

एन एस आई सी
NSIC
ISO 9001 : 2008

CII
Confederation of Indian Industry

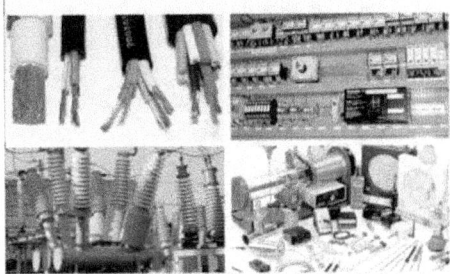

5th
GUJARAT MANUFACTURING SHOW 2014
19 -20 September 2014, The Grand Bhagwati, Ahmedabad

: SPECIAL FOCUS :

Electrical, Electronic & Mechanical Equipment

Accelerating Growth of MSMEs

Event Highlights:

- ◆ **2 Days Exhibition**
- ◆ **Buyer Support Programme (One-on-one business meeting)**
- ◆ Great opportunity to bring together suppliers and buyers to find their business partners. Meetings take place by pre-arranged schedule. Buyers are provided with free meeting table.
- ◆ **Vendor Development Programme**

Exhibition Focus :

The event will feature a focused display of :

- **Power transmission & distribution equipment & technology.**
- **Power dispatching system (SCADA/EMS systems)**
- **Industrial automation**
- **Electrical Energy conversion, distribution & storage**
  - Transformers
  - UPS technologies
  - Rectifier/ Converter systems & equipment
  - Power switches, load switches and busbars
  - PCB Manufacturers
- **Testing, measurement & control equipment**
  - Measurement & automation systems for power engineering
  - Remote control systems
- **Power Transmission & conduction**
  - Non insulated conductors
  - Power cables & fittings
  - Cable lying, storage & processing equipment

- **Power electronics**
  - Batteries & switching regulator ICs
  - Electromechanical & mechanical components
- **Building electrical & Building system technology**
  - Fuses, safety switches, earthing materials
  - Meter boards, series installation equipment, distribution systems
  - Cables and lines
  - Engg. Projects Generator
  - Winding Wires
  - Motors
  - Insulator
  - Stamping & lamination
  - Conductors
  - Energy Meters
  - Capacitors & Registers
- **Mechanical Equipments**
  - Centrifugal fans
  - Bulk material handing conveyors
  - Heat exchangers & condensers etc.

Benefits of Participation :

Brainstorming	**Disseminating:**	**Learning:**	**Building:**
Present Challenges, opportunities and future roadmap	Brand building environment, product demonstration	Best Practices & latest technologies	Partnership & networking amongst all stakeholders

For further details, Please contact
Confederation of Indian Industry
Western Region - Gujarat State Office, CII House, Gulbai Tekra Road, Near Panchwati, Ahmedabad - 380 006
Tel(D). (079) 40279930 I Fax. (079) 40279999 • Board line No. : (079) 40279900-10

Conference
Raja Bhattacharjee Email : raja.bhattacharjee@cii.in
Abhoy Bhattacharya Email : abhoy.bhattacharya@cii.in

Exhibition
Varun Gopalakrishnan Email : varun.gopalakrishnan@cii.in
Ashish Mehta Email : ashish.mehta@cii.in
Anant Jogi Email : anant.jogi@cii.in

Rooftop solar PV system: Will it be a game changer?

The increasing viability of rooftop solar PV systems and their multiple benefits could be a game changer for the country. Although the adoption rate of rooftop solar PV systems is still at a very negligible scale, today, more and more business and corporate houses are adopting this viable and sustainable solution

By Srabani Sen

Over a long period of time, the focus was only on large-scale grid-connected solar power plants. But today, the increasing viability of rooftop solar PV systems, and their multiple benefits, could turn out to be a game changer for the country. Although the adoption rate of rooftop solar PV systems is still at a very negligible scale, the draft guidelines of the second phase of the Jawaharlal Nehru National Solar Mission (JNNSM) has set a target of 1000 MW for rooftop projects, both at off-grid and grid connected levels, during the period 2013-2017. This seems viable in the context of the falling prices of solar (PV) modules. Apart from residences, more and more business and corporate houses are adopting this viable and sustainable solution to cut energy costs and reduce the country's dependence on fossil fuels.

Why should one install a rooftop solar PV system?

Electricity is not only becoming more expensive with each passing day, but also scarce. As the demand-supply gap for electricity is increasing, people are adopting DG sets, which pollute the environment. Hence, it is the responsibility of corporate houses and business enterprises to adopt rooftop PV systems, and become self-reliant and environment-friendly. "Many corpo-

rate houses in India are shifting from conventional energy to renewable energy sources to protect themselves against the continuous increases in electricity costs. Besides the economic benefits given by the government, the other benefits include green energy adoption, corporate sustainability and brand image enhancement," points out Dr Arul Shanmugasundram, executive VP (projects) and CTO, Tata Power Solar.

Says Praveen K Jain, associate vice president, Knowledge Centre, Su-Kam, "Solar PV systems demonstrate environmental consciousness. In the context of increasing energy costs, they help users in significantly cutting down on their electricity bills. Rooftop solar plants offer several advantages over other sources of power, like lower electricity bills, a decrease in transmission and distribution losses, low gestation time, and

Dr Arul Shanmugasundram, executive VP (projects) and CTO, Tata Power Solar

Jasmeet Khurana, senior manager, consulting, Bridge To India

Praveen K Jain, associate VP, Knowledge Centre, Su-Kam

improved tail-end grid voltages."

Praveen Jain also points out that, currently, solar energy is one of the cheapest sources of energy and will become even cheaper in the coming years. The payback period for a rooftop solar PV system is now three to five years, depending upon the type of installation, after which users can derive returns and benefits from the system for the next 25-30 years. In case a corporate house wants to install a rooftop solar PV system, it needs to take approval from the chief electrical inspector to government (CEIG) for connecting power to the in-house grid. For external connectivity, the state's transmis-

As per the latest market studies, India's installed capacity for solar power generation has increased tremendously during the last few years. Rising electricity demand, high irradiation levels, favourable government policies, and huge investments are supporting growth in the solar industry.

The market for rooftop and distributed solar systems in India currently stands at a cumulative capacity of 190 MW. This is almost equally divided between large scale rooftop solar installations (industrial and commercial) and small scale rooftop solar systems (residential and telecom). "The installations in 2013 were down from the 2012 levels as there have been issues with the disbursement of subsidies from the Ministry of New and Renewable Energy (MNRE). Most industry players are now asking for the subsidy mechanism to be removed altogether as uncertainties with the mechanism are causing more harm to the market than good," says Jasmeet Khurana, senior manager, consulting, Bridge To India, a solar energy consulting firm.

"2014 promises to be a year of recovery for the solar sector in India. Bulk of the projects under JNNSM Phase II policy will be constructed through next year and state-led projects are also expected to see some action in 2014. More number of programmes will get funds while PV demand will also increase. Phase II of JNNSM now specifies that 50 per cent of the projects (375MW) need to be built with domestically manufactured cells and modules. Similar provisions such as domestic content requirement (DCR) in state policies as well as future solar policies will tremendously boost local manufacturing and reduce forex outflows and associated volatility in pricing," explains Dr Arul Shanmugasundram.

Outlook by 2018

Bridge To India expects that the rooftop and distributed solar market in India will grow at a CAGR of about 66 per cent, and by 2018, the rooftop solar market will have a cumulative market size of over 2100 MW. "This will include over 750 MW for large scale rooftop systems, more than 450 MW for telecom towers and around 900 MW for residential purposes," adds Jasmeet Khurana.

---

sion company's approval is needed, informs Dr Arul Shanmugasundram.

Types of rooftop solar PV systems

Rooftop solar PV systems are of three types, and users can choose as per their requirements:

- **Grid-tied** rooftop systems supply generated power to the grid and also power the load. A major limitation of this system is that it does not generate power during a power failure as the inverter shuts down the system and no power is sent to the grid.
- **Grid-interactive** rooftop systems work with a battery backup or a diesel generator to support the load. These can work during a power failure.
- **Off-grid** systems work only with a battery backup or diesel generator in off-grid applications.

Components of rooftop system and their capabilities

A rooftop solar PV plant requires only a few components—PV modules (also called panels), an inverter, battery/batteries, mounting structures and a charge controller.

"Solar power conditioning unit (PCU) is an integrated system consisting of a solar charge controller, inverter and a grid charger. It provides the facility to charge the battery bank either through solar power, the grid or a DG set. The PCU always gives preference to the solar power and will use the grid/DG power only when the solar power/battery charger is insufficient to meet the load requirements," explains Praveen Jain.

PV modules: The PV modules could be thin film or crystalline, but experts recommend crystalline panels because these are more efficient for installations where space is a constraint. Thin film modules require more installation space, for the same capacity, than crystalline modules.

Panel efficiency is calculated as per the area it occupies. The capacity of a solar panel, that is, the output of the plant, reduces at temperatures above 25°C and increases at temperatures below 25°C.

Inverters: The inverter and the battery are very important components of rooftop solar systems. They determine the quality of AC power and also the kind of loads that can be powered with solar energy. However, inverters need to be replaced during the lifetime of the plant.

Not all rooftop solar PV plants generate power during power failures. If the inverter uses another source of power as a reference voltage, it can function, but if it is designed to use only grid power as a reference voltage, then the inverter will not be able to function in the absence of grid power and the solar plant will not generate power during power failure.

So one can choose a grid-tied inverter, but it will not generate power during a power failure. An off-grid inverter works only with a battery backup or diesel generator in off-grid applications, and is suitable for applications where grid power is not available, but is not the right choice if one needs a solar plant to

MNRE AND STATE INITIATIVES IN PROMOTING ROOFTOP SOLAR INSTALLATIONS

MNRE has launched a pilot scheme for solar systems that range in size from 100 kW to 500 kW. The Solar Energy Corporation of India (SECI) is the nodal agency for the programme, which aims to generate feedback and further promotion of the concept in the country. Under this programme, rooftop solar systems are to be connected to the grid without battery backup. The surplus power, after consumption in the building, will be sent to the grid. Under the scheme, 30 per cent of the cost would be provided as a subsidy and 70 per cent is to be met by the consumer.

States' initiatives

States are also playing a major role in promoting rooftop solar PV installations. Besides the MNRE scheme, there are several states that have announced their own initiatives for rooftop solar systems, and are coming out with separate plans.

In *Gujarat,* Gandhinagar has seen generation of 1.39 MW of solar power through rooftop installations. Another scheme has also been launched to develop 25 MW of power from rooftops in five other cities.

Kerala has announced the '10,000 rooftops' programme that aims to generate a total of 10 MW of solar power annually by installing solar panels of 1 kW capacity on 10,000 houses across the state. It offers a 20 per cent subsidy over and above MNRE's subsidy. Over 6000 systems of 1 kWp each have been installed in Kerala under this scheme, and around 4000 more will be installed by the end of 2014. Kerala has also announced a similar policy for 25,000 additional rooftop systems.

Haryana is targeting commercial and industrial units for setting up rooftop solar systems to overcome the shortage of power in the state.

Tamil Nadu aims to install 350 MW of rooftop capacity in three phases – a scheme that began in 2012 and will go on till 2015.

Karnataka has a target of installing rooftop solar systems of 5-10 kW capacity on 25,000 roofs across the state.

Andhra Pradesh launched a campaign to encourage house owners to install solar panels on their rooftops and sell surplus power to the state.

Uttarakhand has come up with a policy to harness solar energy by installing solar panels on rooftops and wastelands around buildings.

---

work along with grid supply. On the other hand, a grid-interactive inverter works both with the grid supply and with either a battery backup or diesel generator, and can work during a power failure as well.

A new type of hybrid inverter can automatically manage between two or more different sources of power (grid, diesel or solar). It has inbuilt charge controllers, MPPT controllers, DC and AC disconnects, etc.

Mounting fixtures: Solar panels are mounted on iron fixtures so that they can withstand the wind and the weight of panels. The panels are mounted facing south in the Northern Hemisphere and north in the Southern Hemisphere for maximum power tracking. They are tilted at an angle equal to the latitude of that location.

Trackers: Trackers help to mount the panels in such a way that these panels follow the sun as it moves across the sky. They can increase the power output from the PV plant, but add significantly to both the initial cost of the plant and maintenance expenditure.

Battery: A battery backup can ensure that the load gets sufficient power. Batteries and their charging equipment are not 100 per cent efficient. There is a loss of energy both while charging and discharging the battery. A battery pack can add about 25-30 per cent to the initial system costs of a rooftop PV solar system. Usually, a set of two batteries is installed for each kVA of a solar inverter's capacity.

Charge controller: A charge controller regulates the DC power output from the rooftop solar panels that are used to charge the batteries. It provides optimum charging current, and protects the batteries from overcharging.

Space and capacity required, and the electricity generated

According to Praveen Jain, approximately 7.4-9.3 sq m of roof space is required for a typical 1 KWp solar PV system. A proportionately larger area is required for higher capacity systems. A shade-free roof area is a must since shadows affect the PV plant's performance.

So, how much electricity does a rooftop solar PV system generate? Says Praveen Jain, "Solar energy generation is subject to the solar irradiance (power per unit area) of the location where the system is installed. On an average, a 1 KWp solar PV system can generate 90-150 units of electricity per month."

And how does one ascertain the capacity of the solar power system one requires for a home or office? Says Praveen Jain, "For 3-5 units (depending on the type of solar installation—on-grid or off-grid) of daily electricity consumption of any household, one requires to install a 1 KWp solar system."

Dr Arul Shanmugasundram points out another advantage of installing solar rooftop systems. "If a user has excess electricity, he can inject it from his rooftop installation into the grid through a power purchase agreement (PPA) with the local distribution utility in his area. Under this agreement, a tariff is determined by the appropriate

State Electricity Regulatory Commission (SERC)," he says.

Cost of a rooftop solar PV system

The main hindrance to installing a rooftop solar PV system, apart from the lack of awareness, is the high upfront cost involved. The price range in India is much higher compared to developed markets such as Germany and Japan. The major cost is for the solar panels, housing fixtures, inverters, metering equipment, cables and wiring gear, and batteries. To this, one needs to add the operating and maintenance costs.

According to Praveen Jain, a hybrid rooftop solar PV system starts from Rs 20,000 for a capacity of 850 VA hups with 80 W panel. Su-Kam Brainy is one such solar PV hybrid system. A 1kWp solar PV system starts from about Rs 90,000 onwards, depending on the type of solar installation—on-grid or off-grid.

"The cost of a rooftop solar PV system depends on the function it serves (to feed power into the grid, to support the load during a power failure, etc) and the incentives/subsidies available. It should be noted that all solar PV systems function by matching the voltage from some other source. Therefore, the system has to be integrated with the grid, a battery backup, or a diesel generator," he says.

Though PV module prices have decreased significantly, they account for only half the total cost of the rooftop plant. So any further decrease in panel prices will affect only that portion of the cost of the project. Since the prices of the other components have not decreased the way the price of PV modules has decreased, one cannot expect too much of a reduction in the project costs.

Subsidies that matter

Government provides capital subsidies and tax benefits for putting up a solar

COST OF THE COMPONENTS OF A ROOFTOP SOLAR PV SYSTEM

Component	Rs	% of total cost
PV modules (crystalline)	52,000	52%
Inverters	23,000	23%
Other parts of the system (cables, etc)	17,000	17%
Installation	8,000	8%
Total	1,00,000	

Please note: The above prices are for components from Tier 1 manufacturers that offer 5 years' warranty. Battery backup has not been added as that can alter the calculations significantly.
Source: Solar Mango

FINAL COST OF 1 KW ROOFTOP SOLAR PV PLANT AFTER SUBSIDY

Item	Rs
Cost of a 1 kW rooftop solar plant (estimated)	100,000
Subsidy @ 30%	30,000
Net cost after subsidy	70,000
Accelerated depreciation @ 80%	56,000
Tax rate	35%
Tax saved through depreciation	19,600
Net cost after both A+D and subsidy	50,400

Source: Solar Mango

rooftop system. However, the long delay in getting approvals for projects and in getting the subsidy act as a deterrent.

Capital subsidy: Government provides a 30 per cent subsidy on capital expenditure for rooftop solar PV systems. For commercial and non-commercial entities in grid-connected areas, the subsidy is granted for plant sizes of up to 100 KW. However, entities setting up solar plants for rural electrification can claim a subsidy for plant sizes of up to 250 KW.

Interest subsidy: Government also provides soft loans at 5 per cent per annum on 50 per cent of the capex for a five-year tenure for solar projects by both commercial and non-commercial entities. Commercial entities can claim either capital or interest subsidies. But a non-commercial entity can claim both subsidies, simultaneously.

Accelerated depreciation: For solar PV systems, a company can claim 80 per cent depreciation in the first year, leading to savings on income tax on overall profit. This benefit can be claimed by both commercial and non-commercial entities.

Process of claiming financial incentives: The financial incentives mentioned above can be availed by filling in the prescribed application form and sending it to the Ministry of New and Renewable Energy (MNRE) for project approval. A commercial entity has to indicate its preference for a capital or an interest subsidy. Once approved, in case of an interest subsidy, MNRE forwards the application to a commercial bank for the soft loan. In case the capital subsidy option is selected, MNRE provides the subsidy money in three phases—at the start of the project, mid-way through it and after a successful inspection, post-completion. ▣

THE LATEST IN SMT PRINTER SYSTEMS

A solder paste screen printer for SMT machinery is required to screen print solder paste onto the PCB before placing the surface mount components. Solder paste printing systems are available in three configurations—manual, semi-automatic and fully automatic. The machine can be table mounted, standalone or inline. Many semi-automatic printers offer the manual vision alignment capability, while fully automatic printers offer automatic vision alignment. Some manufacturers prefer stencils over screens because of better image accuracy, volume control and longer service life. Here are some of the latest SMT printers that are available in the market.

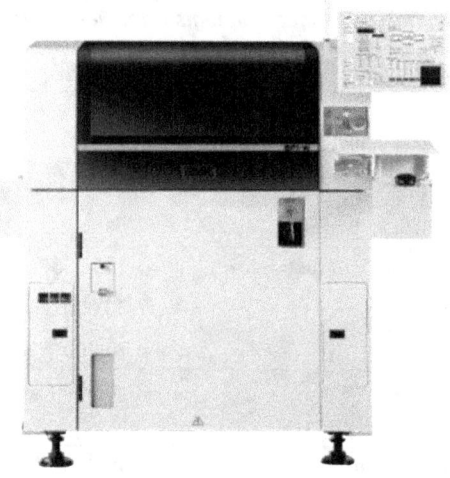

Model: SP1 W, Brand: Samsung, Manufacturer: Samsung, South Korea

Launched in April 2014, the SP1 W screen printer offers an increased print area for large board applications while maintaining accuracy specifications. With its high print quality and reduced cleaning time, the printer increases productivity, maximising throughput and ROI since it is available in a dual lane architecture. It can also allow mixed production of two different PCB models at the same time.

Key features
- Supports PCBs up to 510 × 460 mm; Multiple stencil sizes
- 2D barcode tracking function; Automatic solder paste supply
- Supports single, dual and extension conveyors
- Automatic print height levelling and stencil position setting
- One-touch squeegee replacement.
- Mixed production of different types of PCBs—applicable PCBs: 510 (L) mm × 460 (W) mm (single lane); 330 (L) mm × 310 (W) mm (dual lane)
- Automatic stencil position setting
- Printing accuracy: 12.5μm@6 , Printing cycle speed: 7 sec
- Stencil size (max): 736 (L) mm × 736 (W) mm; Minimum: 550 (L) mm × 650 (W) mm

Contact details: *Sihyun Jayden Jang, global project manager (SMT business), Ph: +82-02-2145-2934, jayden.jang@samsung.com, www.samsungsmt.com*

India branch: *Samsung C&T Corporation India Pvt Ltd*

Model: Gemini, Brand: DEK, Manufacturer: DEK, UK

Launched in April 2014, this printing system comes with the base Horizon iX platform configuration with new added options that deliver exceptional performance and value. Standard fit resources deliver a level of precision that guarantees 6-Sigma performance on every Horizon iX printer of > 2 CPK machine alignment capability @ +/-12.5 microns, and > 2 CPK process alignment capability @ +/-15 microns or better.

Key features
- Temperature and humidity sensor
- Blue under-stencil cleaner
- Semi-automatic stencil load
- Squeegee control feedback
- Tooling deviation monitor
- Advance SPC software (QC Calc) *(options)
- Motion control using CAN BUS network
- Core cycle time: 8 secs
- Maximum print area : 510 mm (X) × 508.5 mm (Y)
- Machine alignment capability: >2 CPK @ +/- 12.5μm, 6 Sigma
- Process alignment capability: >2 CPK @ +/- 15μm, 6 Sigma
- Print speed: 2 mm/sec to 300 mm/sec

Contact details: *Khoon-Heng Lim, sales manager – EMD, Ph: +601-2431-9029, klim@dek.com, www.dek.com*

India distributor: *Maxim SMT Technologies Pvt Ltd*

 **YAMAHA**

JUST FIT SOLUTION No.1

YS12P LED Pick & Place Machine

LED Application

- Automotive
- Plant Cultivation
- Lamp
- LED BLU
- Room Lighting
- Street Lighting

LED Nozzle

LED handling nozzle
Rubber / Plastic / Lens escape

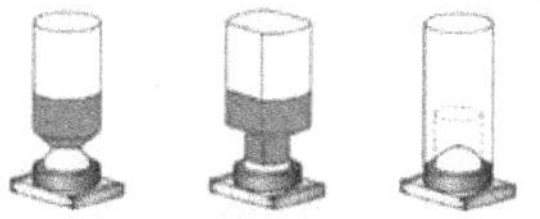

Model	YS12P
Applicable PCB	L50xW50mm to L1,200xW460mm
Mounting accuracy	+/-30um(3σ)
Mounting capability	24,000CPH (0.15sec/chip)
Applicable components	0402(metric base) to 32x32mm Height 6.5mm or less
Number of component types	Tape reel:59 types(8mm width base)
External dimensions	L-1254 x W1440 x H1445mm
Weight	Approx. 1,250kg

24,000CPH (0.15sec/chip)
Excellent Productivity Cost Effective

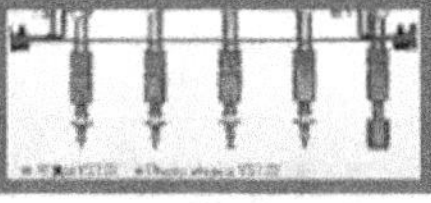

Side-view recognition
Detect upside-down, bring-back of LED

Long-board Handling
Max.L1,200xW460mm PCB can be handled

Electric "SS Feeder"
Intelligent Feeder Less maintenance and long life

- Local service available from Delhi/NCR, Mumbai, Hyderabad, Pune, Bangalore, Mysore and Chennai
- Level II and Level III parts available from within India

 **TRANS-TEC**

DYNAMIC SOLUTIONS FOR THE ELECTRONICS MANUFACTURING INDUSTRY
Transtechnology India Pvt. Ltd.
Plot No.721, Udyog Vihar, Phase- V,
Gurgaon-122016 (Haryana)

Contact No. :-
Ph.: +91 124 6460131
Mobile: + 91 9810449898

E-mail:- trans-tec.india@trans-tec.com
Website:- http://www.trans-tec.com

Model: Yamaha YCP10, Brand: Yamaha, Manufacturer: Yamaha, Japan

Launched in April 2014, the Yamaha YCP10 stencil printer uses a stencil vacuum for a sustainable alignment accuracy of 5 microns along with the correction function to avoid solder filling errors. It comes with the graphical layer base alignment function that is very useful for inconsistent PCB land patterns, therefore eliminating the need for manual training.

Key features

- 5 microns alignment repeatability at 3 Sigma
- 13.5 second cycle time
- Applicable for PCB sizes up to 510 × 460 mm
- Variable stencil sizes available as standard
- Stencil vacuum holder : Vacuum to secure the stencil per se for guaranteed 5 micron alignment repeatability
- Equipped with graphical layer base alignment fuction enables operator to align stencil with PCBs by graphically overlaying the two images, thus negating the requirement of peeping into the machine for inconsistent PCBs

Contact details: Ichiro Arimoto, leader, Ph: 81-53-460-6100, arimotoi@yamaha-motor.co.jp, http://global. yamaha-motor.com/business/smt/

India distributor: TransTechnology India Pvt Ltd For details, refer to advertisement in this issue. See Ad Index on page 8.

Model: SP1200 LED, Brand: EMS, Manufacturer: EMS Technologies Pvt Ltd, India

Launched in February 2014, the SP1200 LED is the first Indian stencil printer built for solder paste printing PCBs that are up to 1200 mm long. It has been designed and developed mainly for the LED lighting industry where PCBs are quite long—ranging from 600 mm to 1200 mm. Streetlight manufacturers, in particular, will find the printer ideal to get good quality and reliable printing.

Key features

- PLC controlled
- Dual squeegee
- Can be programmed for single or double stroke printing
- Pneumatic separation of stencil and PCB
- Squeegee can be tilted for easy cleaning
- Linear guide for squeegee movement for perfect parallel function
- Maximum PCB size of 1200 mm × 300 mm; can also accommodate smaller PCBs

Contact details: L Sampath, director, Ph: 9371077917, l.sampath@emstonline.com, www.emstonline.com

India distributor: EMST Marketing Pvt Ltd

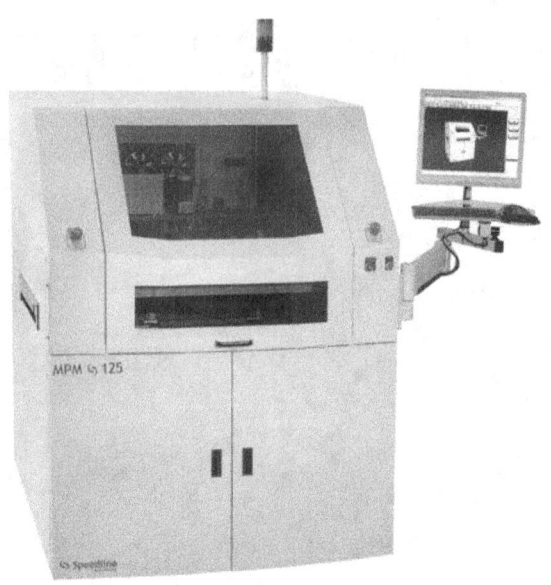

Model: MPM125, Brand: MPM, Manufacturer: Speedline Technologies Inc, USA

Launched in January 2014, the MPM125 screen printer is the perfect union of reliability, capability, flexibility and simplicity. It offers advanced patented technology, high reliability, best in class accuracy and operational simplicity, all in one package.

Key features

- Can handle board sizes of up to 610 mm x 508 mm, board thickness of up to 6 mm and a board weight of up to 4.5 kg
- Vision FOV of 10.6 × 8.0 mm
- Single digital camera with patented look-up/look-down vision technology
- System alignment accuracy of +/-12.5 microns at 6 Sigma,
- CPK of greater than or equal to 2.0

Contact details: *Sam Ong, product manager, Ph: +65-6286-6635, song@speedlinetech.com, www.speedlinetech.com*

India distributor: *Leaptech Corporation*

Model: RP-1, Brand: Juki, Manufacturer: Juki Automations Systems Corporation, Japan

Launched in December 2013, this high speed and high precision solder paste stencil printer comes with the image recognition function and is designed to print cream solder on PWBs (Printed Wire Boards) using screens.

Key features

- Motion screen function for fast and accurate printing. Loading and unloading time is drastically reduced by positioning on the screen side. The alignment stage, where the positioning of the screen is adjusted, has a simple structure that enables high repeatability
- Mark recognition for shape recognition and pattern recognition. The camera switches between the screen position recognition mode and PWB position recognition mode, to allow for quick, accurate position adjustment of the screen and PWB
- Separation of PWB and screen: At snapping off, only the PWB comes down at a controlled speed by the double mechanism at the PWB positioning block. This enables stable printing
- Automatic conveyor width adjustment comes with the machine as a standard function
- The cleaning unit that cleans the back side of the screen can be set in any mode

Contact details: *Akiko Kumon, Ph: +81-42-357-2293, inqsmt@juki.co.jp, www.juki.co.jp/smt_e/index.html*

India distributor: *Juki India Pvt Ltd*
For details, *refer to advertisement in this issue.*
See Ad Index on page 8.

Model: SERIO 4000, Brand: EKRA, Manufacturer: ASYS Group GMBH, Germany

Launched in November 2013, the SERIO 4000 is a unique scalable printing platform with a unique multi-touch interface. SERIO 4000 can also be monitored through tablets, which are used to control ASYS VEGO handling conveyors. SERIO 4000 multi-touch interface called, Simplex, requires no Gerber data for program generation and can handle printing for the latest 03015 size components.

Key features
- Extremely user friendly with a multi-touch interface
- 03015 printing capability with 6 Sigma accuracy
- Capability to print on extra large PCBs
- Line operation monitoring enables checking without being physically present on the production floor
- Closed loop printing controls with on-the-fly feed-back control
- Under one minute NPI introduction
- No Gerber or CAD inputs required for programming

Contact details: *Wolfgang Heinecke, director, sales and service, Ph: + 65 6280 8887, wolfgang.heinecke@ asys.de, www.asysgroup-asia.com*

India distributors: *American Tech, DVS India, EMS Technologies, Innotronics Technologies*

India branch: *ASYS Group India*
For details, *refer to advertisement in this issue. See Ad Index on page 8.*

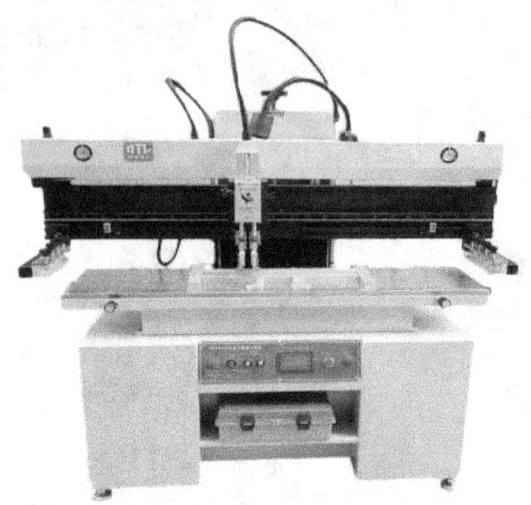

Model: Semi-automatic stencil printer, Brand: Hexi, Manufacturer: Hexi Electronic Equipment Co Ltd, China

Launched in May 2013, this stencil printer comes with a printing platform that has a groove and locating pin, making it easy to set and change the pattern. The printing base can be moved forward and fixed to suit the position of the steel plate, so as to get the best effect. Moreover, speed of printing pressure of the double scrapers can be set by adjusting the precise throttle behind the cylinder to avoid resonance.

Key features
- Scraper can slide back and forth to choose printing position
- Honeycomb board and magnetic thimble that can move arbitrarily
- More applicable to double-sided boards
- Has programmable PLC and human-computer interface screen
- Easy to operate and convenient man-machine interaction
- Emergency alarm for scrubbing steel mesh in time
- Angle, pressure and stroke of scraper can be adjusted; steel scraper and rubber scraper can also be fitted
- Floated scraper system makes printing faster
- Scraper can float up or down freely and can change into horizontal position with steel mesh automatically
- Platform size: 320 × 1300 (mm), PCB size: 250 × 1250 (mm), template size: 550 × 1480 (mm)

Contact details: *Phoebe, international sales, Ph: 0086 755 61517028/ 0086-15889432790, main@hexi-ele. com, www.hexi-ele.com*

India distributor: *None*

   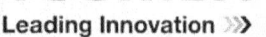 

Today's multimeters are packed with next-gen features

These meters are equipped with capabilities like faster response rates, ability to provide relevant information and more reliable results, besides many other physical attributes

By Richa Chakravarty

Multimeters have evolved into high-end electrical testing equipment. The current market trend and a clear change in customer preferences indicate that the industry is moving towards more advanced multimeters, specifically those equipped with capabilities like faster response rates, the ability to provide relevant information and greater reliability.

A multimeter is mainly used to measure the three basic electrical characteristics of voltage, current and resistance. There are two primary types of multimeters—analogue and digital. Each is designed to measure the same basic electrical values, but differ in their method of measurement and display. The analogue models were developed first, and rely on a moving pointer over a graduated scale to indicate readings. Digital meters use a liquid crystal display (LCD) or light emitting diode (LED) to display readings numerically.

What's new in the market?

Today, most multimeters can measure a wide variety of properties including conductance, resistance, inductance and frequency. Some of them can also measure temperature and duty cycles. The most advanced models have the capability to measure rotations per minute, decibels and microamps. These advanced multimeters have become easy to handle and are more rugged as they have waterproof casings and graphing capabilities. They also come

GW INSTEK'S GDM-834X SERIES BENCH-TOP DMM

- 50,000 counts, VFD display and dual measurement/dual display
- Basic precision of DC voltage: 0.02 per cent
- Selectable measurement speeds and a maximum of 40 readings/s for DCV
- Auto/manual range selection
- True RMS (AC, AC+DC) measurements
- 11 different measurement functions

RISHABH'S 616 DMM

- Provides TRMS measurement, 0.4 per cent basic accuracy
- Tests AC/DC voltage, AC/DC current, resistance, capacitance, frequency, temperature, continuity, diode test and duty cycle
- Dual display, 6600 counts, 66 segments analogue bar graph scale, back light, large display size and other troubleshooting indicators
- Automatic blocking system, CAT IV/III protection class and CAT IV/III compliant test probes

with a variety of accessories including probes, clamps and leads. Some multimeters may come with extra batteries and product warranties that are valid from one to three years. Here are some of the new models in the market.

Meters with a dual display: In the past two decades, the test and measurement (T&M) industry witnessed a shift from analogue to digital technology. Now the shift is towards information-rich displays. Displays used in digital multimeters (DMMs) are undergoing changes. Today, multimeters come with dual display technology which enables

quick diagnosis in real time. "DMMs are evolving to include even more measuring functionality along with better display methodology. The new models include the capability to display two readings such as voltage and temperature, at the same time," informs Prince, product manager, Rishabh Instruments Pvt Ltd.

Fluke's 287 and 289 handheld meters are the next generation meters with high performance industrial logging feature. They have 320 × 240 video graphic array (VGA) display to show multiple readings for easy review of the logged data.

FLUKE'S CNX DMM

- Remote monitoring of circuits from up to 20 metres away
- Data logging of up to 65,000 sets of min/max/average readings using the recording capability
- Flexible recording intervals from 1 second to 1 hour
- Saves time and money by capturing multiple measurements simultaneously
- Helps isolate intermittent events and record signal fluctuations, without users being present, using log function

Chirag Lodhiya, regional sales, electrical division, Ideal Industries India Pvt Ltd

Manish Kwatra, managing director, Metro Electronic Products Ltd

Prashant Jain, senior product manager, electrical, Fluke India

Prince, product manager, Rishabh Instruments Pvt Ltd

Sumit Sharma, marketing manager, India, Good Will Instrument Co Ltd

GW Instek's GDM-834x series, launched in August 2013, is equipped with dual display to support various combinations of measurements. "The result of different measurements simultaneously appears on different displays, which saves users time and the trouble of selecting displays while reading measurement results," says Sumit Sharma, marketing manager, India, Good Will Instrument Co Ltd.

The GDM-834x series also offers three selectable measurement speeds—fast, medium and slow. For instance, DC voltage measurement can reach 40 readings per second on the fast mode, which maximises the effectiveness of each measurement. This mode comes with the convenient USB flash drive storage function, integrated with 12 major measurement items and many auxiliary functions, such as maximum/minimum values, reading hold, relative values, dB, dBm, algorithms and comparison.

In November 2013, Rishabh Instruments launched a new range of multimeters with the dual display system, including the **Rishabh 410, 612, 613, 615** and **616** models. The dual display of these meters provides multiple parameters on the same screen, which helps in faster yet reliable analysis of measurements. The 66-segment analogue bar graph scale visualises the variations in readings, making it more convenient and effective. The display is equipped with higher digit counts for better resolution.

"Engineers require effective visualisation of measurements for better interpretation of results and faster decision making. This is often achieved through in-built displays or advanced computer integration. Our latest multimeters continue to grow in functionality and utility to cover the ever increasing measurement challenges," says Prince.

Wireless/handheld meters: Along with benchtop DMMs, handheld meters are also becoming more powerful for troubleshooting. Customers can today monitor multiple circuits simultaneously on a single DMM, see live trends on a computer or laptop and log the readings at the same time.

Adding to the list of handheld meters, **Fluke** launched the **CNX DMM** in 2013, which is a wireless system that offers more than a traditional multimeter. CNX uses Zigbee radios with a proprietary overlay to communicate wirelessly. Zigbee allows users to connect with up to 10 CNX modules at one time and consumes least power, which maximises battery life. The module battery typically lasts up to 400 hours, which adds up to more than 15 days of continuous recording. It has an inbuilt memory of 65,000 data points, making it an extremely powerful data logging device.

Informs Prashant Jain, senior product manager, electrical, Fluke India, "Fluke CNX is a set of wireless test tools working together. Launched by Fluke in India, with the CNX system, users can take measurements that are remote, simultaneous and recordable. They can take readings and solve problems faster, thereby reducing downtime and increasing productivity."

Metro Electronic Products offers **Mastech's MS8240D** handheld autoranging DMM that provides a USB interface and analysis software. It tests AC/DC voltage and current, resistance, frequency, duty cycle and capacitance, simultaneously.

Meters with advanced computer integration: With the advances in application-specific integrated circuits (ASIC), the capabilities of a multimeter have improved significantly. Multimeters now possess advanced computer integration enabled through RS 232, wireless or Bluetooth technologies. The interface allows the computer to record measurements as they are made. Some

DMMs can store measurements and upload them to a computer, cutting down on time spent and resulting in a higher level of convenience. These DMMs also have features like autoranging, auto polarity, sampling and hold, graphic representation of the selected quantity under test, data acquisition features, etc. "Today's demanding applications need more than just a multimeter. Customers need to log data over a period of time and download it to a PC for further analysis. Fluke provides powerful software—FlukeView Forms which can be used to download data onto PCs/laptops," informs Prashant Jain.

"Multimeters are getting sleeker, even as they become rich in features like incorporating a wider range of parameters in a single instrument. Also, their connectivity to laptops and PCs for data analysis ensures much needed convenience to the user," says Manish Kwatra, managing director, Metro Electronic Products Ltd.

Ideal Industries offers multimeters with advanced computer integration and many features. "Everything that is needed is in the package, including software, USB cable, test leads with alligator clips, a K-type thermocouple and a sturdy protective carrying case. Its high frequency rejection (HFR) mode provides accurate voltage/frequency readings on non-sinusoidal wave forms, such as adjustable speed motor drives," says Chirag Lodhiya, regional sales, electrical division, Ideal Industries India Pvt Ltd.

How to choose the right multimeter

The wide choice available in the market, ranging from lower end basic multimeters to higher end advanced multimeters, can confuse users, making it difficult to select the right multimeter that will meet their requirements and, at the same time, offer the maximum value for the price paid. Here are a few

METRO ELECTRONIC PRODUCTS OFFERS MASTECH'S MS8240D DMM

- Large 22000 counts LCD display with 45-segment bar graph
- Safety rating up to CAT III 1000 V
- Tests AC/DC voltage and current, resistance, frequency, duty cycle and capacitance
- Diode checks and continuity tests
- Autoranging and auto power off, USB interface and analysis software
- Data hold, maximum/minimum and relative measurements

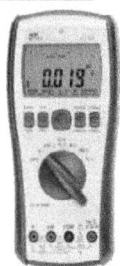

IDEAL INDUSTRIES' 61-498 HANDHELD DIGITAL MULTIMETER

- True RMS readings for error-free testing
- Auto AC/DC voltage and current mode with frequency indication
- Auto selection for resistance/continuity/diode/capacitance
- Large, easy-to-read LCD display, data acquisition, data logging, data storage
- Tests frequency, temperature and duty factor
- Offers dBm/dB measurement and relative mode

parameters users could evaluate before making a final purchase decision.

Bench-top or handheld? Multimeters come in both handheld and bench-mounted variants. Bench-mounted devices are reportedly more accurate than the handheld ones, but a handheld multimeter is more convenient as it can be taken anywhere. So users should be very clear about the type of meter that suits their requirements.

Accurate measurements: Accuracy is the first parameter one should check in a meter before buying it. For example, if you want to discriminate between a circuit that is working poorly from one that is working perfectly, you will need a meter with a higher accuracy level than what low end meters offer. Also, if accuracy matters a lot to buyers, they should go in for a multimeter that can be calibrated. They must also check whether annual calibration is required and, if so, the cost and turnaround time involved.

Display: One should spend some time to define one's display requirements. Essential parameters to consider here are display size, display count, backlight, analogue bar graph scale and other trouble-shooting indicators.

Safety features: Before purchasing a multimeter, buyers must ensure that it complies with international safety norms like IEC 61010-1 and is suitable for the desired protection class (CAT IV/III/II/I). This applies not only to the multimeter but also to probes and any other accessories that come with the device.

Speciality features: The user must also look for features like capacitance measurement, temperature measurement, frequency measurement, and more advanced features like minimum-maximum record, data logging, trending, harmonic ratio, voltage sensing and voltage rating.

Physical attributes: Users should check the physical attributes as well, like size, ruggedness, weight, warranty, battery life—all of which govern the overall life of the multimeter.

THE LATEST IN MULTIMETERS

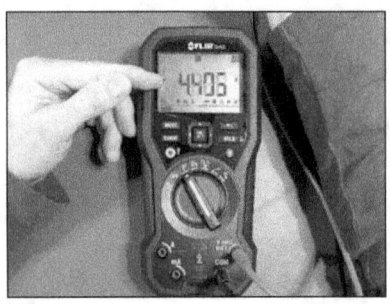

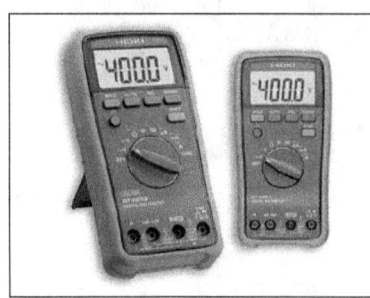

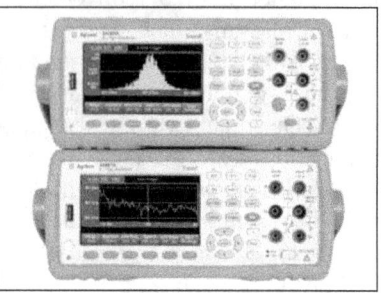

Model: DM93, Brand: Flir, Manufacturer: Flir Systems Inc

Launched in October 2013, the FLIR DM93 is a world-class digital multimeter with advanced VFD filtering to help in accurately analysing non-sinusoidal waveforms and the noisy signals found in VFD-controlled equipment.

Key features

- Uses low impedance input filter to eliminate ghost voltage readings in long-running applications
- Continuous data recording
- Bright dual-LED work lights for dimly lit inspection areas
- Bluetooth connectivity
- Rugged and shock-resistant
- No need to select AC or DC Mode for quick tests
- Backlight features user-configurable timeout
- Large, sharp display is easy to see
- Tool-less battery changing
- Built-in tilt stand

Contact details: flir@flir.com, www.flir.com

India branch: FLIR Systems India Pvt Ltd

Models: DT4211 and DT4212, Brand: Hioki, Manufacturer: Hioki E E Corporation, Japan

Launched in September 2013, the DT4211 and DT4212 models are budget digital multimeters mainly for technicians and educational institutions. The DT4212 comes with the true RMS measurement capability that allows it to accurately measure even distorted current values, enabling usage even in an inadequate power supply environment.

Key features

- Shared features
- Extensive measurement functionality to test multiple parameters
- Large screen for superior usability
- Function that keeps batteries from going dead during measurement
- Operating temperature range of -10°C to 50°C
- Compliance with CATIII 600 V and CATII 1000 V measurement safety standards
- Extensive probes and options to accommodate diverse measurement needs

Contact details: Ph: +81-268-28-0562, os-com@hioki.co.jp, www.hioki.com

India branch: Hioki India Pvt Ltd

Models: 34460A and 34461A, Brand: Agilent Technologies, Manufacturer: Agilent Technologies Inc

Launched in June 2013, the 34460A and 34461A from the Truevolt series of next-generation 6½ digit multimeters help engineers to see measurement data in new ways, get actionable information faster and document their results more easily. Exclusive Truevolt technology reduces extraneous factors such as noise, injected current and input bias current for increased measurement confidence.

Key features

- Offers a basic entry point to the 6½ digit class of DMMs
- Offers expanded current ranges from 100 µA to 10 A
- Comes with temperature measurement function (RTD/PT100, thermistor)
- Expanded diode measurement capability to measure a larger full-scale voltage (5 V)

Contact details: Ph: 1408-3458194, amy_flores@agilent.com, www.agilent.com

India branch: Agilent Technologies Pvt Ltd

Home/office inverters:
What's new in the market

Solar powered inverters and hybrid inverters are currently popular in the market. Typically, these are in the 500 VA to 2000 VA (2 kVA) range

By Kartiki Negi

A wide range of home/office inverters are available in the market today, the result of constantly evolving new technologies. While household customers usually want inverters with a capacity of 150 VA to 2000 VA (2 kVA), in offices, the preference is for inverters in the 800 VA to 2kVA range. These inverters come in pure sine wave, square wave and quasi sine wave. The latest versions come with digital signal processing (DSP) technology, which is the advanced version of the microcontroller based technology. It helps to enhance accuracy, efficiency and protection levels.

The latest entry into the market is the solar powered inverter, which reduces the dependence on mains power, making the product energy and cost efficient. Inverters with multiple battery charging modes and the option to select the type of battery are most popular today.

Another current trend is that UPS systems and inverters are getting merged, eliminating the need for two separate power backup units for powering home and office appliances or IT equipment. Let's review some of the new entrants in the market and the technological advancements they feature.

Solar powered inverters

Launched in December 2013, Ge-

GENUS POWER INFRASTRUCTURES' SOLAR HYBRID INVERTERS

- Delivers 100 per cent pure sine wave
- Ensures round-the-clock protection to the appliances
- Protection from overload, short circuits, high temperature
- AISC (auto sense intelligent control circuit) technology protects the battery
- Ideal for household applications

SU-KAM POWER SYSTEMS' FALCON+

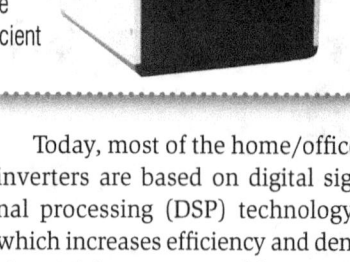

- Can run large appliances like washing machines, water pumps, etc, that require a heavy load
- Compatible charger with all types of batteries-SMF, flooded, LA/tubular
- Increased battery back-up and battery life
- Comes with SMPS charger for fast and efficient charging with continual electricity savings

nus Power Infrastructures offers solar hybrid inverters ranging from 600 VA to 5 kVA. These are integrated with an in-built 10-20 Amp solar charge controller that enables the conversion of solar energy to electricity. "This hybrid inverter ensures maximum use of solar power, resulting in reduced electricity bills and increased energy efficiency. The inverter is also suitable for small and medium business and commercial establishments," says Rajiv Satoor, president, Genus Power Infrastructures Ltd.

Today, most of the home/office inverters are based on digital signal processing (DSP) technology, which increases efficiency and density and improves total harmonic distortion. Genus Power Infrastructures' solar hybrid inverter is also based on DSP technology.

This solar hybrid inverter also has auto sense intelligent control circuit (ASIC) technology, which protects the battery and smartly regulates the charging current, leading to longer life of the battery.

Launched in November 2013, upsINVERTER.com's Sun Pack

SPEED   SECURITY   SELF RELIANCE

Transforming the Value Chain... Realising Opportunities

ORGANISED BY

ELCINA 🔷

ELCINA Electronic Industries
Association of India

EVENT HIGHLIGHTS

- Two days Exhibition
- Conference
- Buyer - Seller Meet
- Special Business Promotion Sessions

FOCUS SEGMENTS

Military - Land Systems
Aerospace - Air force, Avionics
Naval Systems
Homeland Security

"India is set to undertake one of the largest equipment procurement cycles in the world with an estimated spend of about USD 112 billion on capital acquisitions by the year 2016, which will create offset opportunities for the domestic industry worth USD 30 billion"*

(*source KPMG Report)

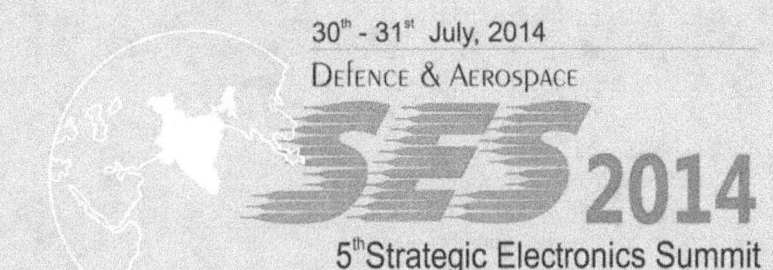

30th - 31st July, 2014

DEFENCE & AEROSPACE

SES2014

5th Strategic Electronics Summit

Bangalore International Exhibition Center,(BIEC), Bengaluru , INDIA

Conference Theme - "Make Indian- Dream to Reality"
Transforming the Indian Strategic Electronics Ecosystem

DAY - I

The Road to Indigenization

- Role of Policies in Indigenization – DPP
- Challenges of Technology Transfer for Indigenization
  - Gaps in Technology & Potential Sources

Regulatory Issues

- Offsets - Implementation & Regulation
- Challenges faced by SME's

Proposed Report on
"Systematic Analysis of Electronics Content in Various Defence Programs"

DAY - II

Defence & Aerospace Market Potential

- Key Defence Procurement Programs with High Electronics Content
- Defence Offsets and their Business Potential
- Capabilities of Indian Industry in Defence Electronics

Silver Sponsor

सी-डॉट
C-DOT

Media Partner

EFY
SINCERELY YOURS

Supporting Media

GLOBAL
SMT & PACKAGING

Segment Partner
Homeland Security

Supporting Association
IACC
INDO-AMERICAN
CHAMBER OF COMMERCE

CEO Networking Dinner Sponsor

ELECTRONIC
MANUFACTURING CLUSTER
SPECIAL INVESTMENT REGION,

GMR

Registration: To Register as a Delegate and / Or book a stall please contact:-

Rajesh Rawat: Mobile: +91 9911445890 Tel: +91 011 41615985 Email: rajesh@elcina.com Fax: +91 011 26929440

ELCINA Electronic Industries Association of India

ELCINA House, 422 Okhla Ind. Estate, New Delhi, INDIA - 110020 URL *www.elcina.com* or *www.sourceindia-electronics.com*

MICROTEK INTERNATIONAL'S 24×7 HYBRID SERIES

- Delivers equally high performance with all types of batteries
- Unique USS (Ultra Sonic Switching) ensures absolutely silent operations
- Pure sine wave output with very low THD under all battery and load conditions
- Bypass switch enables users to isolate inverters from the mains whenever needed

CONVERGENCE POWER SYSTEMS' CPS HYBYX

- Pure sine wave output allows it to run all electrical applications
- Built in, maximum power point tracking (MPPT) solar charge controller (SCC)
- Maximum extended power backup when connected with the correct number of batteries
- Field tested as per Indian conditions

Rajiv Satoor, president, Genus Power Infrastructures Ltd

Sreekumar, director-sales and marketing, Convergence Power Systems Pvt Ltd

Manoj Jain, VP, Microtek International Pvt Ltd

P K Jain, associate VP, Knowledge Center, Su-Kam Power Systems Ltd

Yogesh Dua, director, upsINVERTER.com

home/office inverter uses both solar power and grid power to charge the batteries. However, solar power is the first priority while charging and discharging. "It depends on how customers want to use this inverter-- either as a solar home inverter or as a conventional one. This inverter comes at a low price in comparison to other brands," says Yogesh Dua, director, upsINVERTER.com.

Convergence Power Systems Pvt Ltd launched the CPS Hybyx, a solar hybrid home/office inverter, in October 2013. It can be used for domestic and SOHO applications like lighting, fans, computers, etc.

CPS Hybyx is also a DSP-controlled pure sine wave inverter, equipped with a maximum power point tracking (MPPT) charge controller along with a grid charging system. "The built-in, MPPT solar charge controller enables maximum usage of solar energy," says Sreekumar, director-sales and marketing, Convergence

Power Systems Pvt Ltd.

Sine wave inverters

Su-Kam Power Systems Ltd launched the Falcon + 1050 VA sine wave inverter in March 2014. It sports a contemporary design that matches with the aesthetics of a home. A user-friendly display maks it easy for the user to understand the parameters like battery charge level, room temperature, etc. "Falcon comes with the latest function of multiple battery charging options. Its application areas include lights, fans, TVs, computers, etc," says P K Jain, associate vice president, Knowledge Center, Su-Kam Power Systems Ltd.

Falcon + is equipped with automatic temperature compensation (ATC) technology, which can sense temperature around and accordingly optimise the battery charging function. "This feature helps to enhance the battery life and also increases the battery backup time," adds P K Jain.

Hybrid inverters

Microtek International's latest launch in January 2014 is an inverter-cum-UPS hybrid series, which is a combination of both digital and sine wave inverters. Ranging from 725 VA to 1650 VA, it comes with a switch to select the battery type and works with all the types of batteries available in the market. "It has a switch to select the speed of charging current—standard (10 A) and fast charging (14 A). This keeps the battery ready for frequent emergencies. It also ensures noiseless performance and quick charging, and enables higher load, longer battery life and longer backups," says Manoj Jain, VP, Microtek International Pvt Ltd.

This hybrid inverter is based on

intelligent power saving technology, which makes the product more energy efficient. "With the use of more high-grade components, this inverter saves a lot of electricity compared to other inverters. This technology makes the inverter cost effective and buyers can recover their cost within a span of three years," says Manoj Jain. This inverter also uses intelli battery gravity management (IBGM) technology that increases battery life and performance by maintaining the correct battery gravity.

How to choose the right inverter

The market is flooded with a variety of home/office inverters. The most important factor a consumer should look for is the efficiency of the inverter in terms of evaluating battery performance against costs. Here are certain parameters that would help a consumer to choose the right inverter.

Load requirement: One must determine the application and load requirement before zeroing in on any home inverter. For better efficiency, one should look for the apt capacity required for the setup. It is important to calculate the power consumption, which can be done by adding up the watts (W) of all loads (lights, TV, etc) to be powered by the inverter.

"Buyers must evaluate their total backup requirements as well as the duration of time for which they need the backup, in order to get the best inverter," says Manoj Jain.

Inverter capacity: Next, it is important to select the volt-ampere (VA) rating of the inverter (VA = watts x power factor). Power factor values vary from 0.6 to 0.8. For example, a 600 VA rated inverter (with a power factor of 0.8) delivers approximately 480

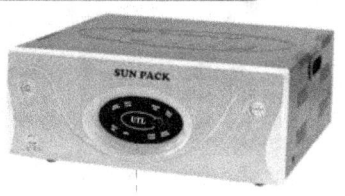

UPSINVERTER.COM'S (UTL) SUN PACK HOME INVERTER

- In-built solar functionality
- First preference to solar power
- Energy efficient
- Charges batteries using solar power

LUMINOUS POWER TECHNOLOGIES' ZELIO

- Grid power–pure sine wave electricity supply
- Intuitive and easy-to-use interface
- Powered by 32-bit DSP processor
- Hassle-free battery maintenance enabled by a digital electrolyte level indicator
- Complete protection for connected appliances
- Wide range of batteries supported

watts only.

Backup time: It is also very important that the backup time of the inverter matches the load of the appliances. Backup time of the inverter can be calculated as Ah × 12 V × PF × 0.9/load VA hours.

Technology: This is the prime factor affecting the performance of an inverter. One is advised to buy a sine wave inverter rather than a square wave or digital inverter, as it doesn't get too heated up, resulting in increased longevity of the appliance. Moreover, it reduces the humming sound that tubelights and fans make—a common problem with square wave or digital inverters. The quality of electricity supplied by a sine wave inverter is also better. "Today, one can also opt for a solar powered inverter, which is good for the environment and saves electricity. It reduces our dependence on the electricity grid for charging. It is also easy to maintain," says P K Jain.

Energy efficiency: One should always opt for an energy efficient inverter with low electricity consumption from the grid. "The consumer should look for intelli power saving technology, as it uses high-grade components that help to save much more electricity compared to normal inverters," says Manoj Jain.

Battery capacity: It is very important to choose a good quality battery and one must be aware of the battery's capacity in order to ascertain the backup time of the inverter. "We offer an inverter with a battery selection slider switch, which allows users the benefit of choosing any type of battery that is compatible with the inverter – be it a normal lead acid or a tubular battery," says P K Jain.

Brand: Consumers should always go in for well known brands, especially in the case of the battery of the inverter. Branded products come with after sales services and a warranty. "Never decide based on just the low price when you are looking for a good inverter. A very high or very low price will not indicate the quality and efficiency of a good inverter," says Sreekumar. EB

Biometric devices are more practical and affordable

With innovations in the field of biometrics, manufacturers are integrating biometric technology with several other security devices to offer fool-proof security

By Richa Chakravarty

According to a Frost & Sullivan report, the biometrics market in India is expected to grow at a CAGR of 48 per cent to touch US$ 359 million by 2016. Biometric technology became popular in India with the UIDAI project. Today, the government and businesses are increasingly adopting biometric devices, as these have become more practical and affordable.

Devices to recognise fingerprints, the iris, retina, face and hand geometry are now commonly used. Signature verification and voice verification devices are also in widespread use. Here is a sneak peek at some of the new biometric devices available in the country.

What's new in the market

With technological advancements, manufacturers are integrating biometric technology with several other security devices to offer fool-proof security. Keeping the consumers' requirements and the changing trends in mind, players are offering complete biometric systems. For instance, advanced devices integrate the face identification feature with an inbuilt camera to capture the images of the employees while marking their attendance.

Also, the latest cloud technology helps users to save huge amounts of money by spending on static IP that synchronises data with a central server in a remote location.

Fingerprint identification: The fingerprint biometric devices used for time attendance at work places are

eSSL'S X990 BIOMETRIC TIME ATTENDANCE AND ACCESS CONTROL SOLUTION

- Push data technology • Users: 3000 fingerprints • Card capacity: 10,000
- Transactions: 300,000 logs
- Sensor: 500 DPI optical sensor
- CPU: 800 MHz processor, 32-bit
- Display: 7.112 cm TFT colour screen

FORTUNA IMPEX'S FACE RECOGNITION-BASED ACCESS CONTROL

- Advanced DSP technology that keeps processing and matching right on the device
- Accurate and fast identification using dual sensor facial recognition algorithm
- Capacity to store 1400 faces (1:N mode) and 1,50,000 transactions. TCP/IP port for communication, built in proximity card reader, USB host for pen drive interface
- User-friendly design for face positioning, contact less authentication and voice prompt
- Multi-authentication methods—facial, RFID cards, etc.
- Low power consumption

among the most popular and common biometric devices across industries, including office complexes, manufacturing units, the hospitality sector, etc.

Fingerprint identification devices with particle image velocimetry (PIV) sensors are the new entrants into the market. PIV technology can identify each particle in an image, using which **eSSL Pvt Ltd** launched the X990 time attendance and access control terminal, featuring a new fingerprint algorithm that provides very fast and accurate

identification. It can verify fingerprints within 0.5 seconds. "X990 can be used in multi-factor authentication that supports fingerprint, password and mifare card. It comes with standard TCP/IP, RS-232/485, a USB-host, USB-client, optional Wi-Fi and GPRS connectivity, and battery backup," says Ankit Shrivastava, regional head channel sales, eSSL Pvt Ltd.

Several innovations have taken place in the fingerprint reader field in recent years. For instance, manu-

INTELLIX'S FAC-300 FACE RECOGNITION TERMINAL

- Authentication relies on live face
- 1:N or 1:1 authentication possible
- Supports up to 1200 faces (1:N scenario)
- Verification speed: Two seconds
- Highly secured way to control door entry
- Security can be enhanced by integration with EM locks, turnstiles, RFID smart cards, etc.

MATRIX'S COSEC DOOR PALM VEIN READER

- Built-in Wi-Fi module
- 3G, 4G, GSM and CDMA connectivity
- 20,000 palm templates, with buffer of 100,000 events
- Ethernet and USB interface
- Dot matrix LCD and touch sense keypad
- Supports RFID cards and PINs
- 1:1, 1:N and 1:G verification

Ankit Shrivastava, regional head, channel sales, eSSL Pvt Ltd

Jatin Desai, product manager, security products, Matrix Comsec India Pvt Ltd

Krishna Mohan, product manager, biometric and access control, Fortune Marketing Pvt Ltd

Soumen Ray, general manager, sales and marketing, Fortuna Impex Pte Ltd

facturers are increasingly offering sophisticated mobile fingerprint readers such as sub-dermal fingerprint readers, which read patterns of blood vessels or tissue beneath the fingerprint, making identity management more accurate and secure.

Facial recognition: Recent technology in facial recognition is addressing and overcoming the challenges posed by image quality (lighting, angle, resolution, obstructions). The latest technology being used is 3D facial recognition—individuals can be identified as they walk past the sensor. The technology involves a 3D vision

system, similar to the human optical system. It creates a 3D human face based on the dimensions and measurements taken by the sensor and matches it with the templates available in the database within three to four seconds. Most of the manufacturers integrate the facial recognition feature with time and attendance machines.

Fortuna Impex launched two new biometric models in April 2013. Informs Soumen Ray, general manager, sales and marketing, Fortuna Impex Pte Ltd, "The face recognition based system is the new technology added to our access control and time attendance

solution. This technology offers a high level of accuracy as it uses the most natural biological features, compared to the traditional technologies available in the market. It is hygienic and completely contactless, and the results are unaffected by the viruses around."

Secureye launched its S-B600 time and attendance-cum-access control machine with the face identification facility in May 2014. "It is based on a 600 DPI sensor and auto adopt technology with 360 degree rotational matching. It increases the matching speed of the device faster than other devices in the industry," shares Krishna Mohan, product manager, biometric and access control, Fortune Marketing Pvt Ltd.

Intellix Security Solutions' FAC-300 is an advanced face recognition terminal that uses high quality infrared cameras to scan facial features and to verify identity. It is designed to cater to the demand for a contactless solution, while maintaining its quality, performance and high reliability. Also, its FaceLock-2 electronic door lock can be customised to accommodate dozens of users. It can be remotely operated, as it is easy to grant and revoke access privileges from users' computers.

Palm readers and iris recognition systems: **Matrix Comsec** is all set to launch a palm vein reader (PVR) based door controller for the highest security and to overcome hygiene concerns. It is an IP65 certified touch screen door controller with PoE connectivity. "COSEC door PVR is a highly secure and contactless biometric device that works by reading the vascular pattern of the palm. Being contactless and technologically advanced, it is appropriate for places like hospitals, chemical industries, factories, corporate houses, R&D centres, etc, where security and hygiene cannot be compromised," informs Jatin Desai, product manager, security products, Matrix Comsec India Pvt Ltd.

4G Identity Solutions has introduced eAccess, an integrated

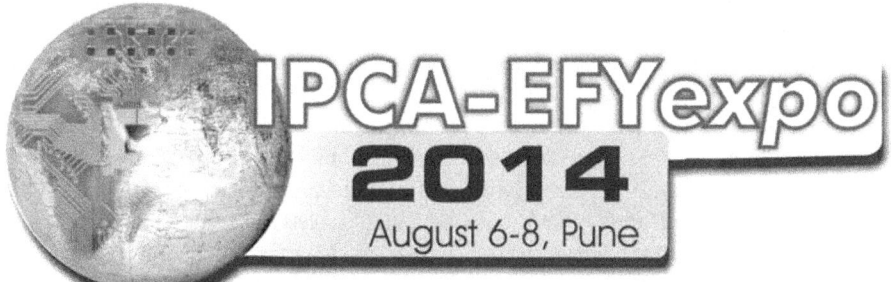

4G IDENTITY'S eACCESS WITH IRIS SCANNER

- Inbuilt storage: Stores over 100,000 users' identity data and more than 100,000 access log transactions within the unit
- Configurable access time duration, card reader interfaces and other inputs and outputs
- Inbuilt LCD display with customisable voice guidance for real-time alerts
- Smart card integration: Integration of mifare and HID iClass smart card systems
- Data redundancy and backup

SECUREYE'S S-B 600 FACE RECOGNITION TERMINAL

- 32 bit high speed microprocessor
- 4.3-inch TFT-QVGA touch screen and touch keypad
- Real-time camera display and online/offline data transmission
- Automatic server synchronisation
- Capacity to record 5000 fingerprints
- Capacity to store, access and manage 5,00,000 attendance records

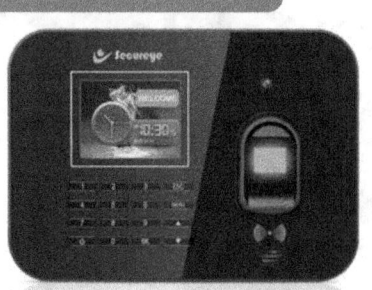

access control system that has been specifically designed to meet the demands of high speed identification and smart access control in highly secure premises. The system is integrated with iris recognition technology that enables real time identification in less than one second, using a database comprising hundreds of thousands of subjects. The best part is that eAccess can be easily integrated with existing security and access control systems of an organisation and is highly scalable.

Logical access control solutions: This is the next big growth area for biometrics, particularly with respect to mobile devices. Today's smartphones can capture voice, face and even hand geometry biometrics without resorting to add-on scanners or other peripherals. In future, tablets can be used for security clearance in organisations, as they can capture enough physical information to authorise the entry of personnel. Today, most of the laptops and smartphones offer biometric features like voice recognition or iris recognition. So, users will not have to use a new password to secure their phones or laptops. Some of the major players in this space include Samsung, Lenovo, Apple, etc.

How to choose the right device

As a wide range of biometric devices are available in the market, a buyer must evaluate the product and the credibility of the manufacturer. The next important factor to keep in mind is the software, which plays an important role. Biometric devices are high end technology products; hence, it is important to know if the manufacturer offers in-house developed software to protect the interests of the buyer.

It is equally important for users to know their requirements as well -- whether they need a multi-functional biometric device (currently used in factories or offices purely for attendance records), or whether a high end sensitive device like a 3D face scanner would serve the purpose better.

The next consideration is the security offered by the device and its supporting systems. As data gathered from biometric devices is sensitive, it should be transferred to a tamper-proof device, transmitted over an encrypted channel and stored in an encrypted database. ▣

CALENDAR OF FORTHCOMING EVENTS

Events	Date	Venue	Organiser
ELCINA Strategic Electronics Summit	**July 30-31, 2014**	**Bangalore International Exhibition Center, Bengaluru**	**ELCINA**
Defence and Aerospace SES 2014 5th Strategic Electronics Summit	July 30-31, 2014	Bangalore International Exhibition Centre, Bengaluru	ELCINA
IPCA-EFY Expo 2014	**August 6-8, 2014**	**Auto Cluster Exhibition Centre, Pune**	**Indian Printed Circuit Association**
ELCINA-EFY Awards	**September 12, 2014**	**India Habitat Centre, New Delhi**	**ELCINA and EFY Group**
NEPCON South China 2014	August 26-28, 2014	Shenzhen Convention & Exhibition Center, China	Reed Exhibitions, Shanghai Branch
ITCN ASIA-IT & TELECOM SHOW	**August 26-28, 2014**	**Karachi Expo Centre, Pakistan**	**Ecommerce Gateway Pakisan (Pvt) Ltd**

The events in bold are the ones EB will participate in

JULY 2014

LED
BAZAAR

**Driving Growth of LED
Business in India**

Innovations in LED technology ■
and wider adoption are opening
up new markets

Buy lumen, not power ■

In conversation with ■

Abhijit R Vaish,
executive director, Instapower Ltd

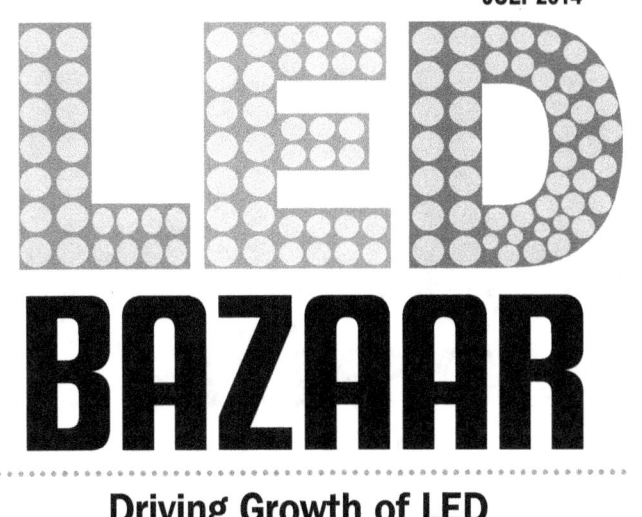

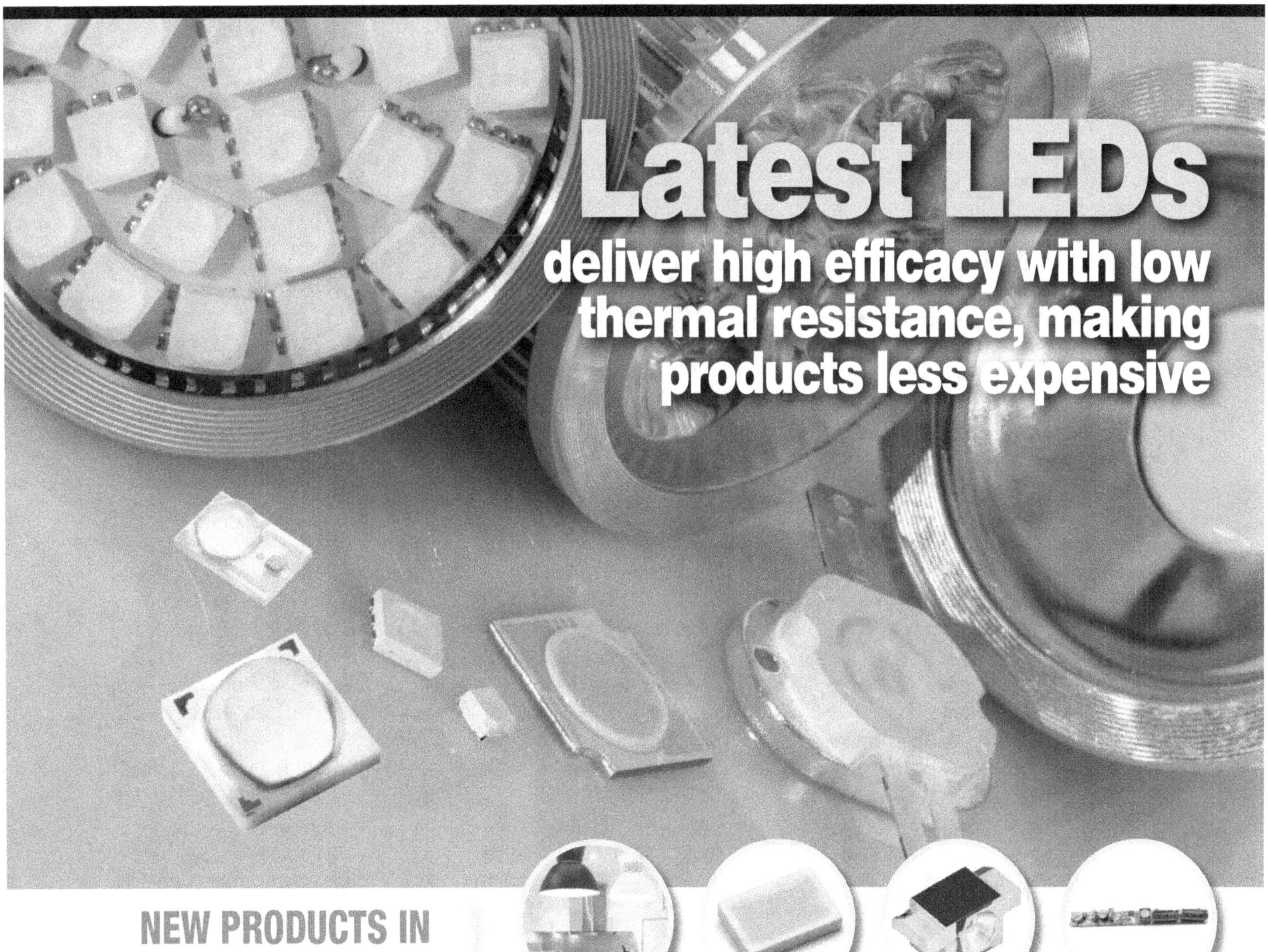

Latest LEDs
deliver high efficacy with low thermal resistance, making products less expensive

**NEW PRODUCTS IN
LEDs, & LED lighting**

Latest LEDs deliver high efficacy with low thermal resistance, making products less expensive

High power LEDs are being squeezed into smaller SMD packages to deliver more power, and thermosetting packaging material is being used to overcome heat dissipation challenges

By Kartiki Negi

High power LEDs, SMDs and lamp LEDs are continuously evolving in terms of luminous efficacy and life span to meet general illumination and backlighting requirements. General applications for these LEDs include streetlights, bay lights, floodlights, etc. In the high power category, both multi-chip and single-chip LEDs are available and offer high efficacy like 160 lumens per watt and above. Let's find out the latest trends in the market.

Latest trends and products

The new trend is for manufacturers to pack more power into smaller SMD packages. Increased use of a chip on board (COB) is the latest development, with which the industry is trying hard to bring more number of LEDs together so as to offer more lumens per watt. "This is a sign of maturity—that the market is ready to experiment beyond the limits that were earlier considered sacrosanct. Today, we have 1 watt in a 5050 PLCC package and 0.5 W in a 2835 PLCC package by incorporating a thermal pad that was earlier absent," says Vijay Kumar Gupta, managing director, Kwality Photonics Pvt Ltd.

In addition to this, many companies are working towards improving the packaging technology of LEDs, particularly to address heat dissipation issues. The latest technology in the high power industry is based on

PHILIPS LUMILEDS' LUXEON LIME LEDS

- Breaks 200 lm/W barrier
- Enables highly efficient colour mixing
- Offers optimal standalone efficiency of 200 lm/W at 350 mA and 85° C
- Provides quality tunable white light in bulbs and fixtures

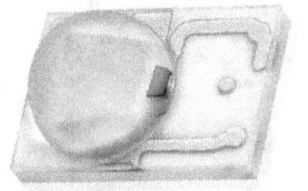

NICHIA'S 219B-V1 HIGH POWER SERIES

- Thermal stability
- Better lumen maintenance
- Offers consistency at Ta85°C

TSLC'S LED HEADLAMP PRODUCTS

- Lumen flux range from 400 lm-1500 lm
- Maximum drive current up to 1000 mA
- Size of emitter: 20 mm x 20 mm
- Offers products with 2-chip, 3-chip, 4-chip and 5-chip versions

thermosetting packaging material instead of standard plastic PPA packaging. As a result, heat dissipation is much better allowing high power 1 W operation, whereas standard plastic packages can only drive in the low to mid-power range of 0.06 W - 0.5 W.

Philips Lumileds has launched a range of LEDs in the last few months. In February 2014, the company launched the LUXEON Rebel ES Lime emitters that offer efficient, tunable white light. Lime enables highly ef-

ficient colour mixing by providing a convenient above blackbody colour point with optimal standalone efficiency of 200 lm/W at 350mA and 85° C. The spectral output of Lime is closely aligned with the wavelength that human eye cones are most sensitive to—555 nm.

In April 2014, Philips Lumileds came up with the new LUXEON CoB 1202 LED arrays, offering 130 lm/W at 85° C. This array, along with compatible reflectors and drivers, enables

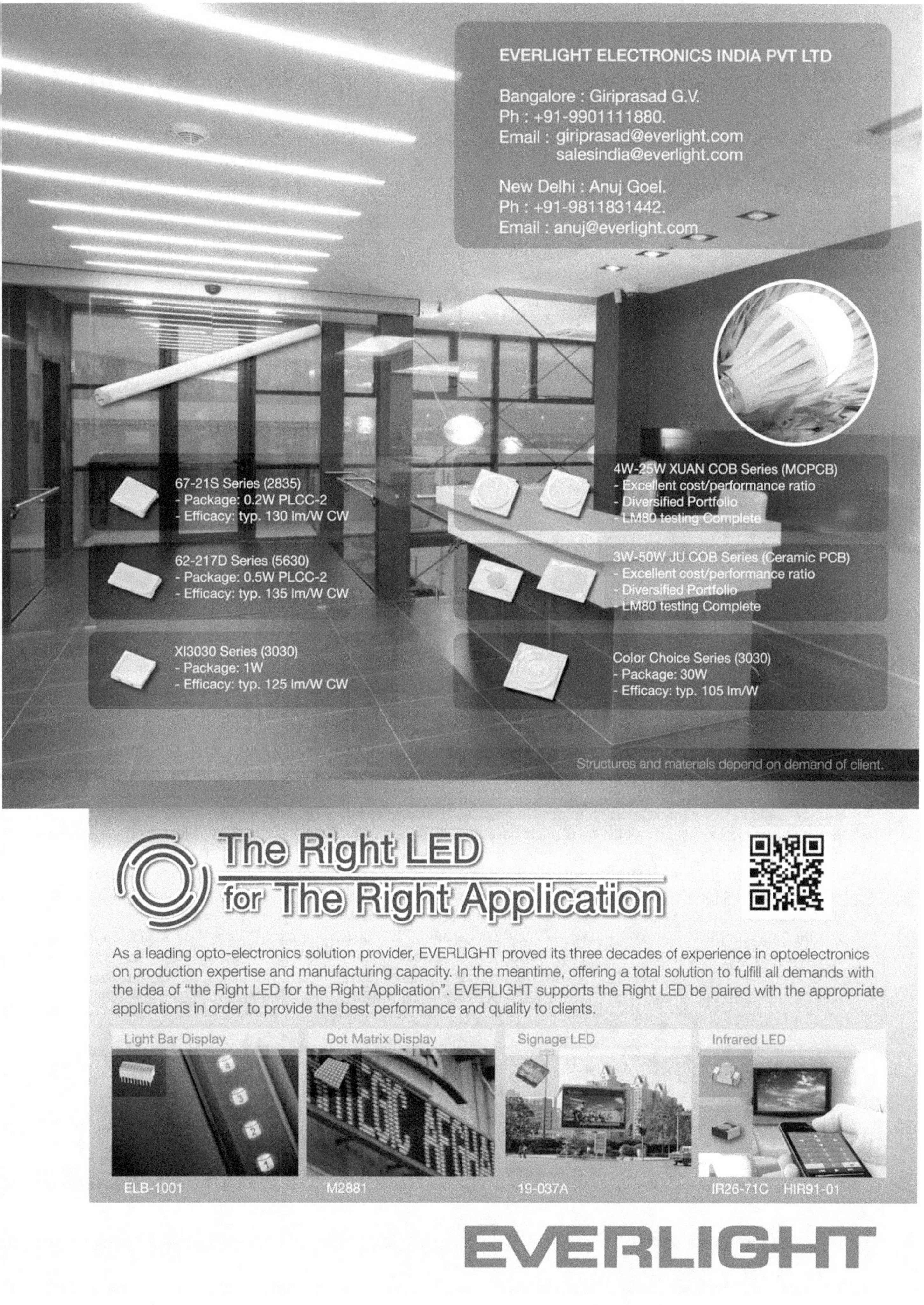
EVERLIGHT ELECTRONICS INDIA PVT LTD
Bangalore : Giriprasad G.V.
Ph : +91-9901111880.
Email : giriprasad@everlight.com
salesindia@everlight.com
New Delhi : Anuj Goel.
Ph : +91-9811831442.
Email : anuj@everlight.com
67-21S Series (2835)
- Package: 0.2W PLCC-2
- Efficacy: typ. 130 lm/W CW
62-217D Series (5630)
- Package: 0.5W PLCC-2
- Efficacy: typ. 135 lm/W CW
XI3030 Series (3030)
- Package: 1W
- Efficacy: typ. 125 lm/W CW
4W-25W XUAN COB Series (MCPCB)
- Excellent cost/performance ratio
- Diversified Portfolio
- LM80 testing Complete
3W-50W JU COB Series (Ceramic PCB)
- Excellent cost/performance ratio
- Diversified Portfolio
- LM80 testing Complete
Color Choice Series (3030)
- Package: 30W
- Efficacy: typ. 105 lm/W
Structures and materials depend on demand of client.
The Right LED
for The Right Application
As a leading opto-electronics solution provider, EVERLIGHT proved its three decades of experience in optoelectronics on production expertise and manufacturing capacity. In the meantime, offering a total solution to fulfill all demands with the idea of "the Right LED for the Right Application". EVERLIGHT supports the Right LED be paired with the appropriate applications in order to provide the best performance and quality to clients.
Light Bar Display
ELB-1001
Dot Matrix Display
M2881
Signage LED
19-037A
Infrared LED
IR26-71C HIR91-01
EVERLIGHT

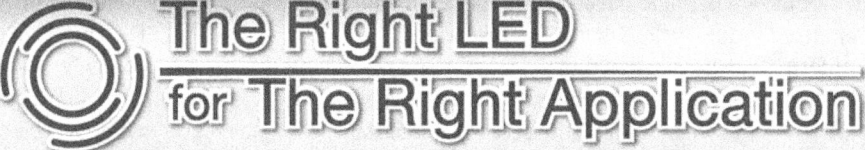

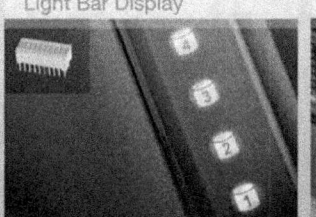

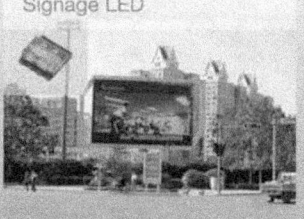

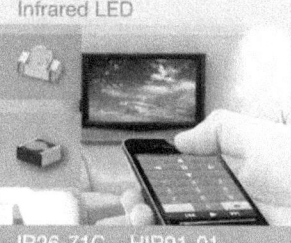

KWALITY'S KLSL505W AND KLSL2835W LEDS

- Incorporated thermal pad which was earlier absent
- High-volume lead frame
- Used for 350 mA and 150 mA

EVERLIGHT'S XI3030 SERIES AND XI3535 SERIES

- Standard 3 V/6 V versions
- High voltage version
- Low CCT 2200 K
- Multiple colour versions

Amrith Prabhu, country manager, Philips Lumileds

Chandra Bhanu, manager-technical LED division, Nichia Chemical India Pvt Ltd

Ewing Liu, technical marketing manager, lighting application section, Everlight Electronics Corporation Ltd

the most affordable downlights, spotlights and PAR38. "The LUXEON array LEDs deliver the highest efficacy available in the market with the lowest thermal resistance. CoBs also have the smallest light emitting surface (LES) with the best uniformity. This enables products with the highest efficacy in the least expensive systems," says Amrith Prabhu, country manager, Philips Lumileds.

The performance range of the LUXEON CoB 1202 arrays is 130 lm/W hot,

over a correlated colour temperature (CCT) range of 2700-5700K at a CRI (colour rendering index) of 70, 80 or > 90. The typical output for warm white (3000K, 80 CRI) is 800 lumens when driven at 200 mA.

Nichia Chemical India Pvt Ltd recently launched the 219B-V1 high power series that delivers 170 lm/W with a maximum driving capacity of 1500 mA with lower forward voltage. It has better thermal stability, better lumen maintenance and consistency at Ta85°C. "Its application areas are streetlights, bay lights and floodlights, to name a few. These are single chip LEDs and have an easy-to-handle secondary lens that is available with all big lens manufacturers. Hence, this gives us freedom to play with different types of light distributions," says Chandra Bhanu, manager- technical, LED division, Nichia Chemical India Pvt Ltd.

Everlight's existing range offers two different packages—XI3030 series

and XI3535 series, using thermosetting packaging material. It comes with multiple variations, in standard 3 V and 6 V, with high performance and high voltage versions for more simple and cost effective driver designs, a super warm 2200 K CCT version to emulate halogen or incandescent lighting options, and individual colour versions for colour mixing or decorative usage.

"The new package is a good balance of high performance at an affordable cost. It is less than 20 per cent of the cost of traditional ceramic high power products. This package platform allows for different technological variations depending on the chips used, colours mixed and driver conditions," says Ewing Liu, technical marketing manager, lighting application section, Everlight Electronics Corporation Ltd.

Launched in November 2013, **Kwality Photonics'** KLSL505W105 Lm 350mA LEDs are a clever combination of optimal chip size and a high-volume lead frame. Earlier, this 5050 package was used only for 60-120mA ratings. Kwality also launched the KLSL2835W 65 Lm 150mA LEDs in the same month. Previously, the 2835 package was limited to 30-60 mA ratings only. "These LEDs allow manufacturers to retain old PCB footprints and yet graduate to higher power without increasing the number of LEDs. The 2835 W goes into bulbs and downlights, while the 505 W is used for low cost downlights and streetlights, without loss of performance," says Vijay Kumar Gupta, managing director, Kwality Photonics.

In April 2014, the **Taiwan Semiconductor Lighting Company** (TSLC)

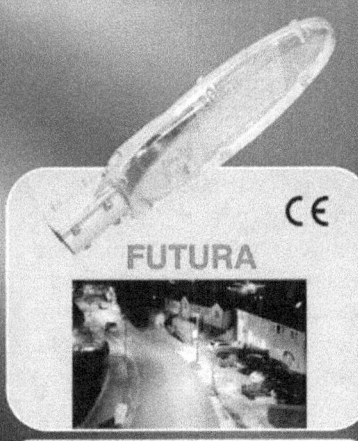
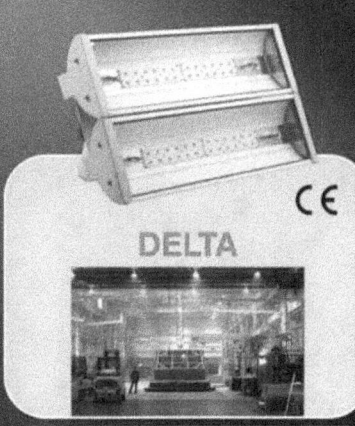

launched high power IR LEDs, the C3535X-Inx1, for surveillance applications. TSLC is developing packaging based on hybrid technology for wavelengths that range from UV lights with invisible short wavelengths to visible light with long wavelengths of blue, green, red and IR. Developed with ceramic package technology, this LED is highly reliable, has high emission power and a long lifespan. "The C3535X-Inx1 has a radiometric power of 700 mW under currents of 1 A and a WPE of over 30 per cent. This can lower system costs even when incorporating all CCTV product features at the same time," says Gopal Shukla, country manaer, SemiLED Taiwan Semiconductor Lighting Company.

Meanwhile, TSLC also launched new LED headlamp products, which are especially suitable for automotive and motorcycle headlight applications due to their excellent and high thermal stability, and high luminance characteristics. Considering the different design requirements of head lamps, TSLC offers a series of head lamp products including 2-chip, 3-chip, 4-chip and 5-chip versions. The total lumen flux can range from 400 lm to 1500 lm, and the maximum drive current can go up to 1000 mA. "The chip used in a head lamp has excellent high temperature stability and very low lumen decay even while operating in high temperatures, so it's very suitable for high temperature applications such as headlamps in automobiles," says Gopal Shukla.

The overall size of the emitter is only 20 mm × 20 mm, so these products are able to operate with a

OSRAM OPTO SEMICONDUCTORS' SYNIOS E4014 SERIES OF MID-POWER LEDS

- Offer low 0.57-mm profile
- 4×1.4-mm footprint
- Colour maintenance at 0.5 W operating levels
- Available in a choice of 4000 K or 5000 K CCT
- Delivers a 120°beam and 41 lm output at 100 mA of drive current
- CRI of 80

Gopal Shukla, country manager, SemiLED Taiwan Semiconductor Lighting Company

Vijay Kumar Gupta, MD, Kwality Photonics Pvt Ltd

second lens such as PES or reflector type and are used in high-beam, low-beam and fog lamps.

How to choose the right LEDs

When looking for suitable high power LEDs, it is very important for consumers to be satisfied with respect to several parameters (e.g., the lumen bin should be on the higher side while for-

ward voltage should be on the lower side). One should always focus on the lumen maintenance and colour stability of the product after 10,000 hours of operation at higher temperatures. Factors such as the level of efficacy, quality of light, heat dissipation and renowned brands should be considered before zeroing in on a product.

Efficacy: Once customers have selected the lumens needed for their application, they must look for the best efficacy levels for LEDs, that is, lumens per watt (lm/W). This will reduce the operational expenditure of the LED fixture. High power LEDs have the flexibility to over drive or under drive to create different ranges of efficiency.

Quality of light: Reliability and quality of light in terms of colour consistency over time are some of the very critical points to consider when buying LEDs. The quality of the LEDs is measured in terms of how long it takes for the initial brightness and colour to deteriorate.

Heat dissipation: When selecting high power LEDs, buyers should ensure that they have enough heat sinking to dissipate the power needed to generate the required lumens from the LEDs.

Optics: Buyers should also consider what type of optics is being used, so as to ensure there are no hot spots and to create the correct light pattern.

Reputed brands: One should choose a renowned brand, which will not compromise on the quality of the LED die, phosphor, encapsulant and leadframe materials, which together determine the quality and performance of an LED. ▣

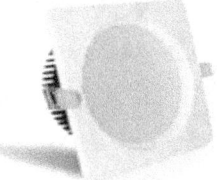

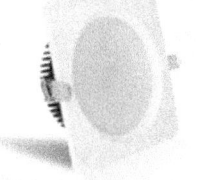

❝With LED luminaires becoming competitive on price & performance, market penetration will increase

Abhijit R Vaish, executive director, Instapower Ltd

Instapower Ltd has been a leading name in the LED lighting space. It is recognised as an R&D house by the Department of Scientific and Industrial Research, Government of India, and has 90 patents to its credit, 10 of which are in the LED lighting and energy efficiency domain.

In conversation with **Srabani Sen** of *Electronics Bazaar,* **Abhijit R Vaish, executive director, Instapower Ltd**, talks about the trends in the market, the challenges being faced by domestic luminaire manufacturers, and how the recent proposal to set up two semiconductor fabs in India can benefit the company.

EB: What are the major trends that are evolving in the LED industry?

The LED lighting industry is evolving fast, with a wider range of products arriving in the market every month. While the initial focus had been on replacement lamps, 2013 saw more innovative designs that had LEDs working with sensors and controls. Another major trend is the constant improvement in the performance of LED lights. And with LED luminaires becoming more competitive on price and performance, their market penetration is bound to increase.

EB: Has the Indian LED manufacturing sector derived any benefits from the recent government policy to boost domestic manufacturing?

The impact of the policies recently rolled out by the Department of Electronics and IT (DeitY) is yet to be seen. Although the policy talks about subsidies for setting up manufacturing facilities, it is too early to gauge whether these subsidies have had any impact in increasing domestic manufacture of luminaires. India still lacks skills development facilities; hence, there is a shortage of trained manpower. And the advanced machinery required for manufacturing still has to be imported. So, hopefully, the recent government policies will be able to resolve these issues.

EB: How would you rate India's expertise in LED manufacturing?

Despite having the expertise, currently we cannot manufacture a lot of things in India. For manufacturing, we require an extremely good tool room, as well as good fabrication facilities for aluminium and die cast parts, both of which are important but are still not manufactured in India. Die-casting of certain components of very specific designs is something we have not been able to do in India, as of now. Second, due to the lack of certain types of machinery in India, many components cannot be manufactured here. Third, the lack of encouraging domestic demand means that in India, we cannot yet get into high volume manufacturing. Without the volumes, one cannot justify the cost of developing new LED products. Countries like Taiwan and China manufacture at a larger scale as they get high volume orders. These are some of the major concerns that hinder manufacturing in India, despite the expertise we have.

EB: As a luminaire manufacturer, what do you expect from the two upcoming fabs?

These will definitely make a huge difference. India is already late in setting up its own fabs, considering that China has nearly a 100.

Currently, only 5 percent of the components we use are manufactured in India; the rest we procure from China, Taiwan and S. Korea. When these fabrication facilities come up in India, we can expect the components industry that earlier existed in India to get revived. We will have a lot of opportunities to manufacture these components in India and that will reduce the cost

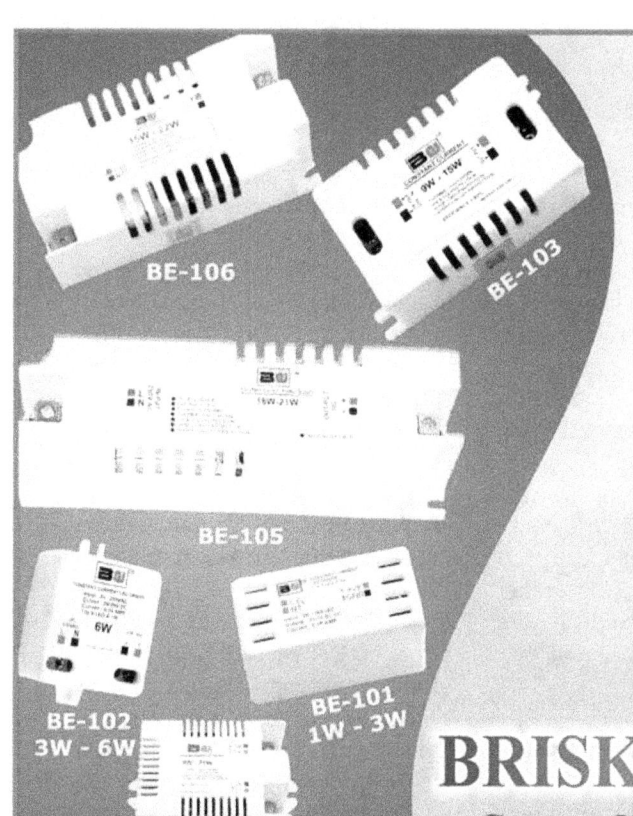

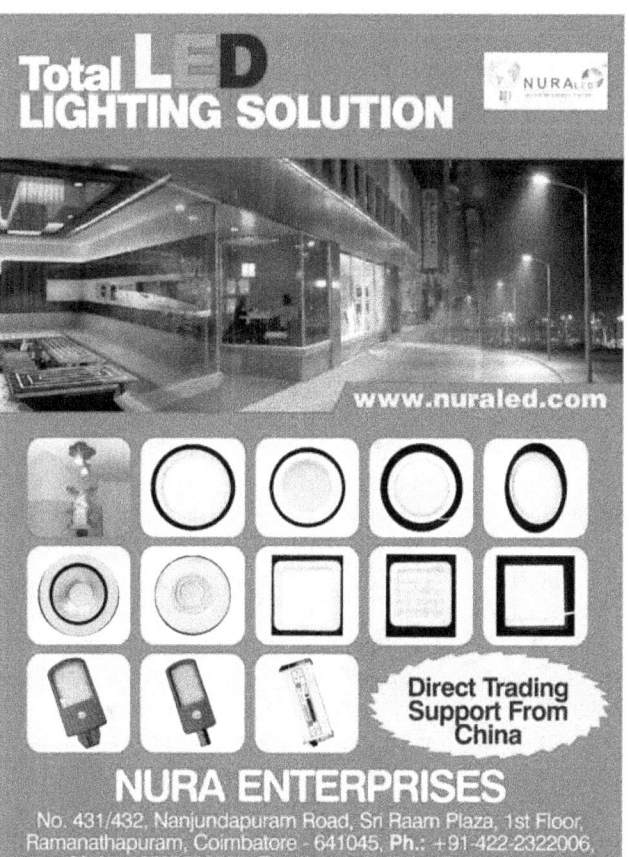

of production. It will be an important step towards curbing our forex outflow. I was reading a report that stated that we have a major forex deficit because most of our money is going to China, South Korea or Germany to buy the raw materials needed for the electronics industry.

Second, the fabs will generate a lot of employment, and hopefully, that will lead to having some skilled hands in India. We need to empower the youth with professional skills, as we lack skilled hands in India. I have personally visited factories in China, Taiwan and Germany, and there is a huge difference in the skills those people possess compared to our workforce. We really have to bridge the skills gap; otherwise, we are heading for trouble.

EB: How is Instapower continuing to manufacture in this scenario?

It is a tough time for any manufacturer in the LED sector, as this sector is still unorganised. We, therefore, face a lot of issues in deliveries, as the supply chain is also very unorganised. Our suppliers, too, face a lot of problems and as a result, they cannot keep up to their commitments and deadlines. Sometimes they face electricity problems, at other times it is labour issues, besides the shortage of raw materials -- the fluctuating prices of raw materials lead to suppliers waiting for the prices to come down. All these add up to the problems that manufacturers face.

EB: What does Instapower manufacture?

We offer complete LED lighting solutions—so we manufacture the luminaire (except the LEDs) and the drivers. We also manufacture LED displays and some solar products.

EB: Where does Instapower's expertise lie?

Our expertise lies in manufacturing the complete luminaire with backward integration. We offer lighting solutions for both indoor and outdoor applications, but we are stronger in indoor applications. We also design our own drivers as per the Indian conditions, factoring in the problems of voltage fluctuation, surges, etc. We are strongest in the aviation lights segment.

We produce everything based on indigenous technology. In fact, we have 90 patents to our credit. Last year, we won the 'Technology leader' award from Frost and Sullivan. Our in-house R&D division is led by a very capable team.

EB: What does innovation mean to Instapower?

Innovation to us means indigenisation. As long as we are able to indigenise our products and customise them as per our local conditions, we will stay ahead of our competitors. For example, in the case of a solar product, to make it work in Rajasthan as well as in Jammu and Kashmir is a big challenge, with the weather conditions being so diverse. We have developed an innovative solar product that works in all kinds of climatic zones. These innovations keep us ahead of competition.

EB: Do you also customise based on the requirements of your clients?

Yes, we do a lot of customisation for our clients. But before committing to that, we consider the volume of the business and the potential of the product.

EB: Does customisation reduce or increase costs?

It works both ways. Sometimes, the customisation is done based on a larger perspective. For example, aviation lights work on AC but our R&D team designed a light that can work with solar energy as well. That customisation was done for a particular customer, but then it became a benchmark for others. It was a one-time cost, which helped to get us volumes. So that customisation was for the good. On the other hand, if the customisation is very specific, then the cost goes up, and ultimately, we need to increase the cost of the product.

EB: What business strategy do you follow to penetrate new markets?

With more awareness about the benefits of LED lights, and with the price going down significantly, penetrating new markets has become easier. We have been very strong in the B2G and B2B sectors. But now we are planning to also enter the retail space, which has opened up as the competition has increased and the acceptance levels have gone up.

EB: Did you launch any new products this year?

We have launched a couple of downlights this year, specifically for the indoor space. We have made these aesthetically pleasing to appeal to customers, and we have also improved upon their functional performance. At the same time, we have kept the price low. We have also launched a dimming solution.

EB: What are you expansion plans?

With our second facility in place, we now have the required bandwidth. So we will not expand our production facilities as of now. However, we are working on building up our market share by creating a network of dealers and distributors. We are working closely with our channel partners and are trying to push our products into the market through sustainable strategies. We are going slow, as being aggressive in a new area may backfire. All these years, we have been working in the B2G and B2B sectors. EB

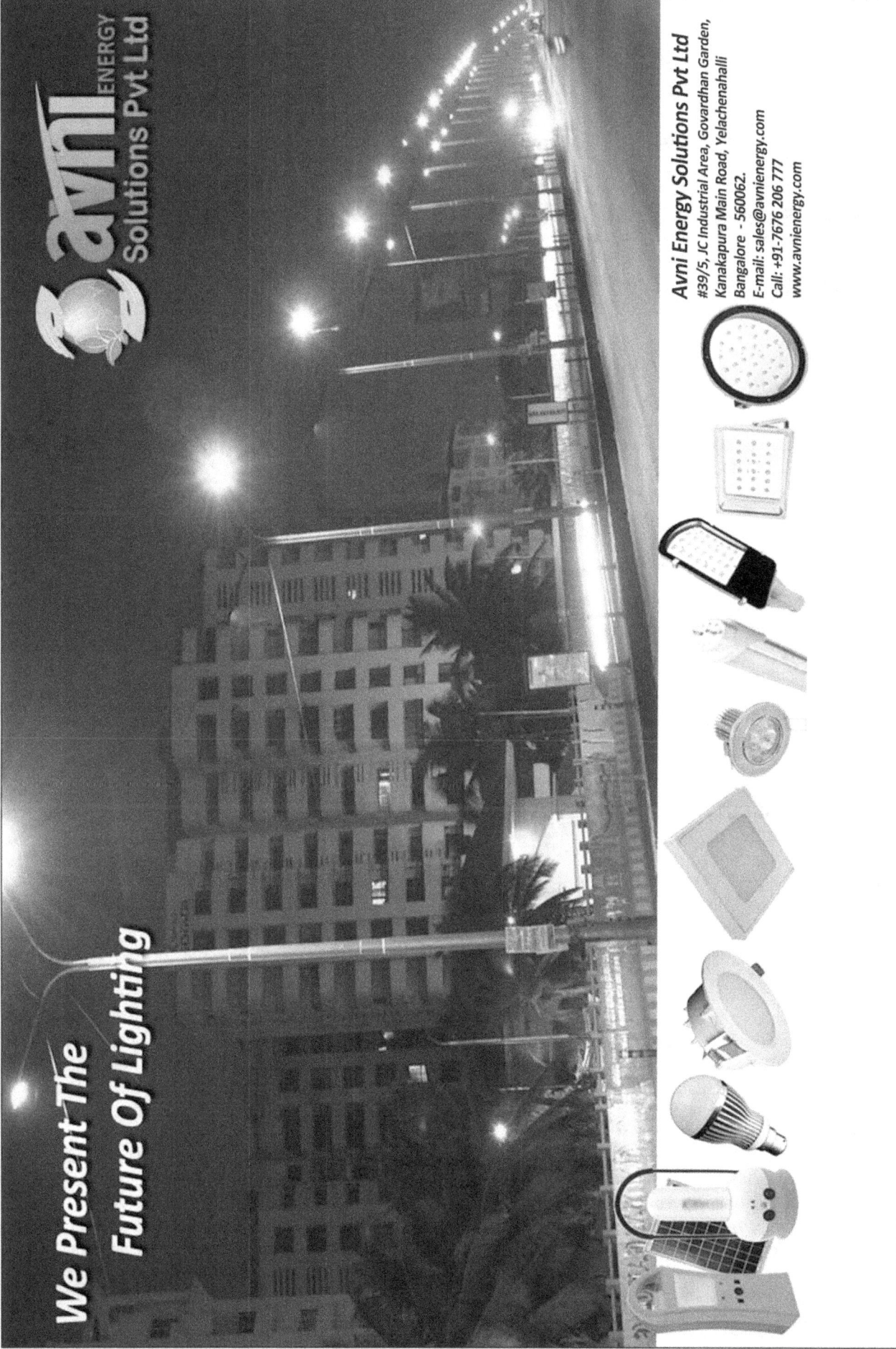

Innovations in LED technology and wider adoption are opening up new markets

The adoption of LED lights in several new sectors is being driven by the latest technologies, expanding the application base for green lighting

By Kartiki Negi

LEDs and LED-based products have been evolving continuously and so has the market. Today, most people are familiar with the benefits of LED bulbs, which was not the case four or five years back. Speaking at the seminar: '*A to Z About Manufacturing LED Products in India*', organised by the EFY Group on February 22, 2014, during the EFY Expo 2014, Amrith Prabhu, country manager, Philips Lumileds, said, "With the continued adoption of LED lighting, we will see more and more markets opening up, as the technology has evolved rapidly, enabling LEDs to be used in more applications."

Indian market scenario

Although traditional energy efficient technologies like compact fluorescent lamps (CFL) and high intensity discharge (HID) lamps are still in the mainstream, it has been forecast that LEDs will take the lead position in India within the next two to three years.

"LEDs in streetlights have been adopted at a fast pace across the country and this area is where LEDS were most widely used. But, in 2013, the industry saw a major transformation taking place in the indoor lighting market. Downlights started picking up as they became affordable, offering significantly improved total cost of ownership (TCO). LED lamps, on the other hand, have not been adopted to the extent they should have been. At present, we can also

Amrith Prabhu *at EFY Expo 2014*

LED Seminar

see the trend of growing LED bulb usage. Therefore, the pace of growth in the Indian market will soon match that of the rest of the world in the coming years," Amrith Prabhu said.

In 2013, the trends in the market showed mid-power LEDs being used in indoor applications. Breakthroughs in mid-power LED products heightened their visibility and triggered growing demand. Commercial lighting, in particular, requires high quality LED products, while for the rest of the lighting market, the emphasis is on ease of use and competitive prices, backed by good quality.

Factors driving demand

Today, adoption of LEDs is quite fast. LED TVs, displays, display backlights in laptops, automotive LEDs, etc, are already very common. The key factors driving the LED illumination market include the high quality of light, the technology being 'green', lower costs and growing awareness about energy efficiency.

Price has been a driving force for LEDs and LED products. In the past, LED lighting costs were very high and unaffordable. However, increased awareness and the latest technology have led to a huge fall in LED prices. This has majorly impacted the adoption of LEDs and LED products.

Apart from price and efficiency, the quality of light is a very critical factor for user acceptance. The light emitted should be realistic, reliable and have predictable performance. Moreover, it should save energy on the system level

with the possibility to fully utilise the 'connected lighting' aspect of LEDs. "When we talk about the quality of light, we stress on the following five elements—colour spectrum and rendering, colour consistency between sources, colour in application, colour consistency in beam and colour consistency over the lifetime," says Amrith Prabhu.

LED technology is opening up new markets

LED lighting technology holds a lot of promise when compared to other sources of light. "The efficiency levels of LED lights will soon go up to 220-250 lumens per watt depending on CCT. Performance is definitely going to increase, with room for improvement in internal quantum efficiency (IQE). And there is scope for efficacy levels to go up further with advances in technology. As of now, we do not see any competing technology, which is affordable, efficient and can replace LED technology. Over and above all this, prices are coming down drastically. With efficacy also increasing, cost reduction will continue in the future, too. So this gives us the confidence that we need to focus on LEDs," says Amrith Prabhu.

At present, LED lights have migrated from the 5 mm radial devices of a few years ago to the high power LEDs, then to mid power versions and on to chip on boards (CoB). LEDs are now application optimised, which means that there are different LEDs for indoor, outdoor and industrial requirements. "For streetlights, you need more than 15,000 lumens, which is definitely different from what is required for retrofit purposes, which would only require 600 lumens," informs Amrith Prabhu.

Recent trends indicate that low power LEDs will eat into the market share of mid power LEDs, which in turn will cannibalise the high power LED market. High power LEDs will continue to be used for application specific needs, and mid-power LEDs will further evolve and deliver better performance. Apart from the high power and the mid power LEDs, CoBs and array technologies have also evolved.

These different trends and technologies are opening up new market segments, which have immense business potential.

Retrofit bulbs: The biggest market across the world is for retrofit lamps or replacement bulbs, which are gaining popularity as a result of the low energy consumption and running costs of LED bulbs. Used in retail, residential, industrial and commercial markets, LED solutions for retrofit bulbs offer a high quality of light by leveraging high power LEDs.

Outdoor lights: The outdoor lighting market is the second biggest market, especially for street lighting. However, here, the major push has been from the government. A majority of the LED streetlight projects in India are funded by the government.

Retail and hospitality: The major reason for LEDs gaining popularity in the retail segment is the fast payback factor, as LED systems are far more efficient than the conventional metal halide lamps. LED lighting for the retail space is more versatile, offers better quality of light and can definitely challenge the popularity of CFL or HID lamps.

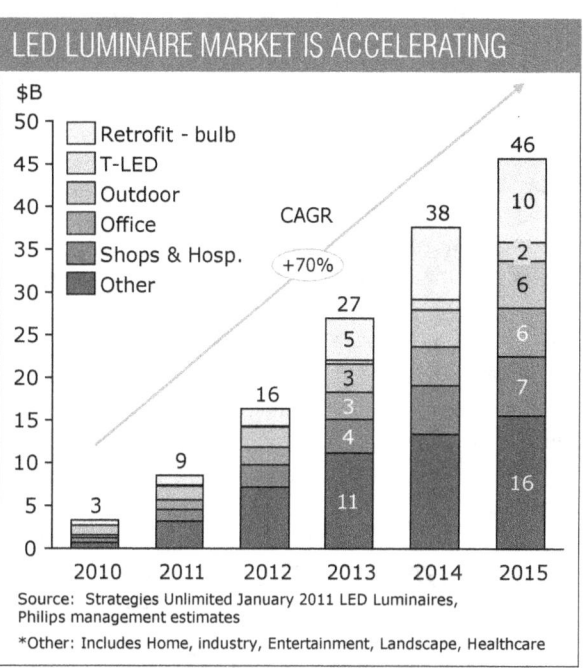

LED LUMINAIRE MARKET IS ACCELERATING

Legend: Retrofit - bulb, T-LED, Outdoor, Office, Shops & Hosp., Other

CAGR +70%

2010: 3
2011: 9
2012: 16
2013: 27 (5, 3, 3, 4, 11)
2014: 38
2015: 46 (10, 2, 6, 6, 7, 16)

Source: Strategies Unlimited January 2011 LED Luminaires, Philips management estimates
*Other: Includes Home, industry, Entertainment, Landscape, Healthcare

LED LIGHTS ARE GOING DIGITAL

We are entering a digital world— from using telephones we've moved to cell phones, from cell phones to smartphones and from notebooks to tablets. Similarly, LED lights are also heading towards 'connected lighting'. We can now control the entire lighting of our homes from our mobile phones; advantages include the ability to remotely dim lights, increase brightness, change colours to set up a particular mood, etc.

Offices: This is the fourth largest segment for LEDs, globally. This segment has picked up very fast because of the improved total cost of ownership, and the exceedingly good quality of light.

Architectural lights: Globally, the architectural sector has adopted LEDs in a big way, though this is not quite true for India. China and Hong Kong are using LED lights for most of their ancient buildings as these lights offer an intensive colour portfolio. This is a very large market.

Industrial market: This is the slowest growing market. However, with the portfolio of LED lights for the industrial sector increasing, it is also gaining traction in applications like industrial high bay, batten and well glass applications. **EB**

Buyers' Zone

Buy lumen, not power: Greenstar

A long lasting luminaire is a combination of the right LED and an incredible heat sink design

By Kartiki Negi

Much has been said about the benefits of LEDs in terms of efficiency and efficacy. Similarly, there should be more awareness about LED luminaires, which will be most efficient and have a longer life if the right kind of LEDs are fixed into the right fixtures and components.

This was emphasised by Greenstar, a group company of Toshiba Lighting & Technology Corporation, Japan, at a seminar 'Little Known Facts About LED Lighting', organised with an aim to create awareness among the LED consumers about advantages of LED lighting and how can it contributes in reducing light pollution.

Several eminent speakers from Greenstar such as Chuk Chakravarty, CTO, Katsuhiro Shinosawa San, CEO, Sarosij Sengupta, director, India technology and operations centre, and Rajat Behal, VP, sales and marketing spoke about the latest LED technology. Among the others were Amrith Prabhu, country manager Philips Lumileds and Murili Devrajalu, senior manager-green initiatives Infosys.

Speaking at the seminar, Sarosij Sengupta said, "Buyers must understand that the life of the luminaire is linked to more than just the LEDs life, which is 60,000 hours." He explained that the luminaire's life depends on three aspects—life of the LEDs, the power supply/electronic driver and the operating environment. Even if one of these fails, the whole luminaire will fail.

Environment conditioning for LEDs

Sarosij Sengupta pointed out that the operating environment of the LED and luminaire is the most important factor influencing performance, since lights perform differently in different environments. Therefore, the luminaire should be made to match the application of a product. "A long lasting luminaire is a combination of the right LED with a 60,000-hour life and an incredible heat sink design, as heat is the biggest enemy of an LED light," said Sengupta.

Choosing the right luminaire

Sengupta pointed out that choosing the right luminaire will lower the total lifetime cost of the lighting installation. "For a good LED light, one should look for total luminous flux, luminous intensity distribution, B-U-G rating, electrical power characteristics, luminous efficacy and colour characteristics (CRI, CCT, etc)," he said.

The seminar concluded that buyers needed to check the following factors in any LED luminaire, in order to ensure it lasts for a long time:
- Adequate illumination, that is, lumens
- High uniformity as there should be no dark patches
- No light should be wasted, so zero uplight
- Quality of light, which includes colour rendering index (CRI), correlated colour temperature (CCT), etc
- Maximum efficacy in terms of light per energy unit

Importance of tests

In a country like India, LED lights are exposed to different environmental conditions such as high and low

Katsuhiro Shinosawa San, CEO, Greenstar, Japan, at the event

temperatures, voltage fluctuations, water and dust pollution, vibration, surges, spikes and thermal shock. An LED luminaire should be protected from all these factors and, hence, should be put through several tests before being launched in the market, said Sengupta. Some of the prominent tests include in-situ temperature measurement tests, which measure the LED source temperature within the LED luminaire or lamp. Another test is the TM-21-11, which is a method of taking LM-80 data and making reasonably accurate lifetime projections of the LED product. The high accelerating light test exposes the LED to temperatures of 300°C to 400°C. If the product passes this test successfully, then it can easily withstand the normal temperature fluctuations.

Users should also be aware of the standards that LED luminaires need to comply with. They must ask for the following protection while buying an LED light:
- Optical performance as per the LM79 standard
- Minimum 3G vibration resistance
- IP66—water and dust ingress protection, etc. ■

Driver IC for LEDs

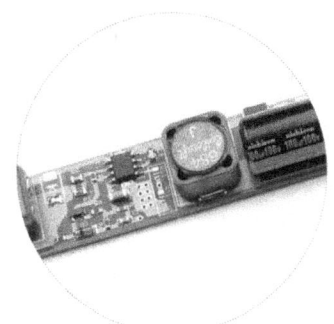

In May 2014, NXP Semiconductors launched the SSL5018, a compact, highly-efficient driver IC for LED solutions. Complementing NXP's broad GreenChip SSL family of products, the SSL5015/18, with an integrated MOSFET, supports lighting applications up to 18 W lamp power like the TLED and PAR lamp shapes. The controller is available across multiple markets. The SSL5018 is designed for Europe and Asia. By introducing it in India, NXP also adds four new reference boards to the NXP LED driver solutions selection guide. NXP's cost-effective LED driver solutions simplify lamp design by using the right form factor to streamline the design-in process, and extend lifetime and reliability.

For further details: *Ph: 080-40240000, www.nxp.com*

Low bay lights

In April 2014, **GlacialLight** introduced the Arcturus series of GL-BL50 low bay lights. These dimmable 50 watt LED lights come in three colours and in an artistic design, making them suitable for a variety of indoor environments. Shopping malls, restaurants, offices and even homes can benefit from the even lighting and contemporary styling of the GL-BL50 series. Offered in a white, silver or black matte finish, a blue ring encircles the GL-BL50, giving off a cool ambient glow when the light is turned on, while the main lighting source projects downwards from the rounded lower skirt. Hanging options include from a chain, pendant rod, or using cable suspension, ensuring these low bay lights fit in any environment.

For further details: *India distributor: Key Operation and Electrocomponents Pvt Ltd, Ph: 08377806486, www.philipslumileds.com*

T-shaped mid-LED

In April 2014, **Osram Opto Semiconductors** launched a compact side-facing T-shaped mid-LED, the SFH 4140, that offers a high radiant intensity and protrudes only 0.6 mm above and below the board, offering the remote control function even within a very small space. It is the company's latest addition to its portfolio of remote controlled transmitters, a no-compromise low-profile solution for which there will almost always be sufficient demand. This infrared transmitter can be integrated in extra-thin smartphones or tablet computers. It takes up only 4.6 mm^2 of board space and emits a powerful focused beam to the side. The beam angle of $+/-25°$ is created by an integrated reflector that is a real plus point in terms of the space requirements.

For further details: *India branch: Osram India Pvt Ltd, Ph: 0124-2383180, www.osram.com*

Mid-power emitter

In March 2014, **Philips Lumileds** introduced its mid-power LED, the LUXEON 3020 emitter. This LED solution will drive a variety of long-awaited commercial LED lighting fixtures, including lamps and troffers, into the mass market. The LUXEON 3020 is the most affordable of all its mid-power LEDs, delivering over 1,000 lumens per dollar. It also features hot colour targeting and a 1/9th micro colour binning structure. Philips Lumileds' hot colour targeting ensures the colour temperature remains within ANSI specifications at operating conditions. For designers of lamps, troffers, TLEDs, high bay and low bay luminaires, the LUXEON 3020 produces 90 lm at 6500 K and 80 CRI when driven at the maximum drive current of 240 mA.

For further details: *India distributor: Key Operation and Electrocomponents Pvt Ltd, Ph: 08377806486, www.philipslumileds.com*

New manufacturing facilities in India

By EB Bureau

The company	About the facility
1 **Honeywell Automation India Ltd** **Manufactures:** Electronic instruments, process control systems, sensors, etc	Honeywell Automation India Ltd has set up its new manufacturing facility approximately 25 kilometres from its main campus in Pune. Spread across 6968 sq m, the facility includes space for electronics manufacturing, warehousing and offices. The company will be manufacturing electronic instruments, process control systems, sensors and process transmitters in this facility *Contact details:* Ph: 91-124-4975003, madhavi.jha@honeywell.com, www.honeywell.com
2 **Swelect Energy Systems Ltd** **Manufactures:** Solar PV modules and converters	Swelect Energy Systems has commissioned its 15 MW solar energy park near Vellakoil in Karur district, Tamil Nadu. The total investment involved in the project is around Rs 1060 million. On an average, 75,000 units of power will be generated by this plant. *Contact details:* Ph: +91-44-24679618, www.swelectes.com
3 **Gautam Solar Pvt Ltd** **Manufactures:** Solar off-grid products	Gautam Solar, formerly known as Gautam Polymers, has set up a new manufacturing facility in Haridwar. Spread across 1000 sq m, it is located near its existing facility with 25 MW of annual capacity. *Contact details:* Ph: +91-9313314063, solarlights@gautampolymers.com, www.gautampolymers.com
4 **Harman International** **Manufactures:** Audio equipment for automotive and consumer electronics	Harman International has set up a production facility in Pune. The company will assemble car infotainment systems for domestic car makers. The Indian facility is expected to go on stream by June 2014. The company has invested around Rs 200 million in this venture and the Pune facility will employ around 100 people. *Contact details:* Ph +91-80-43306451, indiaproinfo@harman.com, www.harman.in
5 **NTPC Ltd** **Deals in:** Power generation (generation and sale of electricity)	NTPC has completed its 50 MW solar power project in Rajgarh in Madhya. With this, NTPC's portfolio of completed solar power plants adds up to a capacity of 95 MW. In March, it announced the commissioning of the Talcher Kaniha (10 MW) solar power project. Another 15 MW solar power plant in Singrauli, Uttar Pradesh, is currently under development. NTPC's objective is to generate 128 GW of power by 2032. Its target is to generate 28 per cent of its power from non-fossil fuels. *Contact details:* www.ntpc.co.in
6 **Welspun Renewable Energy** **Deals in:** Engineering, procurement and construction (EPC)	Welspun Renewable Energy has commissioned a 19 MW solar project in Chitradurga district in Karnataka. While the company had earlier commissioned 8 MW, it took just three months to commission the remaining capacity of 11 MW. This project will supply power to the grid for the next 25 years with enough clean energy to power 48,000 homes. *Contact details:* Ph: +91-11-66034600, welspunenergy@welspun.com, www.welspunenergy.com
7 **Schwing Stetter India** **Manufactures:** Concrete handling equipment	Schwing Stetter India has set up a solar power plant at its manufacturing complex near Chennai. The solar plant has been installed with an investment of Rs 6.5 million to save on power costs. *Contact details:* Ph: 044-27156780/1, chennai@schwingstetterindia.com, www.schwingstetterindia.com

SOUTH SPECIAL

NORDSON INDIA PVT LTD

Suprotik Das, MD

Nordson India carries out the engineering, marketing, manufacturing and customer support activities required by Nordson's customers in South Asia. In addition, Nordson India designs and develops next-generation products and technologies for use in Nordson's markets worldwide

Headquartered in Bengaluru, Nordson India is responsible for managing Nordson's ADS, EFD, ICS and PPS business interests in the seven countries of South Asia: India, Pakistan, Sri Lanka, Bangladesh, Nepal, Bhutan and Maldives.

Today, Nordson India's team members are spread across 11 cities in India—Bengaluru, Ahmedabad, Chandigarh, Chennai, Hyderabad, Kochi, Kolkata, Mumbai, New Delhi, Pune and Rudrapur. In addition, it is also represented in Colombo (Sri Lanka) and Karachi (Pakistan).

Services offered

With 70 direct employees, Nordson India carries out engineering, marketing, manufacturing and customer support activities required by Nordson's customers in South Asia. Nordson India also designs and develops next-generation products and technologies for use in Nordson's markets worldwide, including hardware and software solutions.

Nordson India's Engineering Resources & Technology Centre in Bengaluru houses laboratories with a comprehensive range of test equipment and the capability to simulate production processes in a controlled environment. These labs are used to develop solutions for customers and even for training—of customers as well as Nordson's own personnel. In addition, the Centre designs and develops new products that are uniquely suited for the Indian market and other developing economies.

The most recent development is a powder booth made of composite materials. This is similar to the Apogee material used in the US, but is a good deal less expensive. These booths have given Nordson India a big competitive advantage in the country.

Products offered

Nordson India has also developed innova-

tive automation systems to add value to Nordson's products. Such systems have been successfully delivered for applications like labelling stock, surgical tape, tamper-evident cartons, solar thermal panels, water filters, energy meters and many others. While many of these have been installed in India, Nordson's subsidiaries and distributors in other emerging countries have also purchased Nordson India systems from time to time. As a result, Nordson India equipment can today be found in countries like Saudi Arabia, UAE, etc.

In 2000, Nordson India created its Embedded Systems Group, to develop software and hardware solutions for its global ICS business. Today, this group has grown to 14 people who develop new hardware and software products, test them and even develop low cost sources. Currently, this activity is done for the ICS and ADS business groups. Many of the products developed by this group are already being offered in global markets: the Prodigy manual gun controller, the Vantage automatic controller, the Vantage manual gun controller, iTrax PRx Controller, Encore XT Controller, and many others.

Strengths

The majority of Nordson's employees in India are qualified engineers or managers, and most of them are trained extensively at Nordson's international facilities. This combination of infrastructure, skills and training, at locations that are close to customers in the territory, has resulted in Nordson's ability to offer world-class capabilities to customers in South Asia. This aspect is seen as one of the key factors behind Nordson's success in the region.

Contact details: 143A, Bommasandra Industrial Area, Bengaluru 560 099, India Ph: 91 80 4021 3600, www.nordson.co.in

ACCUREX SOLUTIONS PVT LTD

K N Rammohan, CMD

Accurex offers solutions right from the conceptualisation of an electronics company, to getting on to the drawing board for a building plan, process flow and product flow charts, construction, flooring, procurement of machines, etc

Accurex Solutions Pvt Ltd started in a small way back in 1987, and has since come a long way in fulfilling its promise as an electronics assembly house. Its growth is based on continuous sustainable expansion, consistent after sales service support, and a commitment to be partners in the progress of its customers.

The Accurex group has completed 27 years in the industry, for which all credit is due to its dynamic leader and visionary CMD, K N Rammohan, who has been associated with the Indian electronics industry for more than 30 years. His vast capabilities have enabled him to build an unbeatable sales and service team.

Accurex today has emerged as a group of companies to reckon with, which is not only sensitive to its responsibilities to its clients but also to society.

The company has its head office in Bengaluru, also known as the Silicon Valley of India. Its branches are headed by independent decision makers in Mysore, Hyderabad, Chennai, Mumbai, Jaipur and New Delhi. Accurex also has an overseas office in Singapore.

Services offered

The Accurex group covers PCB assembly equipment, components, embedded solutions, customised ATE, test solutions, etc. Accurex also sells and offers after sales support in some of the SAARC countries.

The family of Accurex employees is always ready to delight its customers through hard work and good support. Accurex is committed to providing after sales support and service to more than 500 clients in the areas of aerospace, defence electronics, medical electronics, consumer electronics, EMS, etc.

Accurex boasts of a dedicated, highly result oriented and well experienced marketing team, which is always on the move in order to support its customers. The company has highly qualified service engineers who attend to customer service calls 24 × 7.

The Accurex group offers solutions right from the conceptualisation of an electronics company, to getting on to the drawing board for a building plan, process flow and product flow charts, construction, flooring, procurement of machines and tools for assembly, testing and packing, sourcing of components and sub-assemblies, building customised testing solutions and simulation of display. Between the various firms in the group, virtually every requirement of the electronics industry is covered.

Tie-ups

Accurex has been working with some principals and OEMs since a very long time. These include Mydata, Reprints, JT Electric, Mirtec, Nordson Dage, Top-A, Pace Inc, Xuron, Armeka, Treston/Sovella, SMT Germany, Glenbrook, Ex-More, Aquesous Technologies and CC Hydrasonic.

Corporate social responsibility

The KNM Education Trust was founded by the Accurex directors as part of its corporate social responsibility. Some of the directors of Accurex give lectures on entrepreneurship and SMT technologies in various institutions, thus sharing their knowledge with the younger generation.

Contact details: No. 782, 1st Cross, 3rd Main, 5th Phase, BEML Layout, Rajarajeshwari Nagar, Bengaluru 560098, Ph: 080-28611871, info@accurexsolutions. com, www.accurexsolutions.com

BORG ENERGY INDIA PVT LTD

Dr Boaz Augustin Jr, CMD

BORG is committed to serve the energy industry and create unique products for the utility market. Its products ensure uninterrupted power supply, the highest efficiency and maximum savings

Incorporated in 2013, BORG Energy India Pvt Ltd is a subsidiary of BORG Inc, Texas, USA. It has rich experience in alternative energy and smart grid technology, having successfully completed various projects across North America, Europe, Africa, China and South East Asia. BORG Inc has an installed capacity of more than 4500 MW providing solar energy solutions to its customers.

The company has earned revenue of over Rs 1 billion in the Indian subcontinent and has established more than 30 franchise outlets across the country; it plans to add another 190 outlets over the next four months. BORG Energy India has also appointed more than 90 distributors and 1000 dealers across India. Its projected revenue for this financial year is Rs 4 billion.

Products offered

BORG offers its Astra Plus Home Series micro solar power plants to household customers to harness solar power. The unique smart grid technology ensures uninterrupted power supply for a sustainable future. BORG's other domestic products include the Farm Master Series and Commercial Series.

These smart grid solar systems are very easy to install and operate. They demonstrate an improved conversion efficiency and higher durability. Moreover, their alternating current coupling technology protects the life of the system and battery, ensuring flexible energy management.

BORG Inc has also launched off-grid PCUs designed for domestic and medium-sized industrial operations in India to replace utility power during emergency shutdowns.

Customer service

Power Play is what the exclusive BORG showrooms are branded as. They feature an attractive display of BORG's domestic products, enabling customers to get a hands-on experience of the products. BORG Care Centres offer the best in-house service network to address the queries of its customers. All BORG products are internationally certified and have an innovative approach to renewable energy generation. as they combine the power of distributed solar energy solutions with smart grid technology.

BORG is committed to serve the energy industry and create unique products for the utility market. Its products ensure uninterrupted power supply, the highest efficiency and maximum savings, and they are the first of their kind in India.

BORG is also planning to set up an R&D facility in 2016 that will focus on nano technology in solar panels. This team will work on several technologies to enhance the recovery of energy, in the white light, ultra-violet and infrared portions of the spectrum.

Contact details: No: 15–16, 6th floor [Opp Trinity Church], Vayudooth Chambers, Mahatma Gandhi Road, Bangalore–560 001, Ph: 080–4947 6600, marcom@borgenergy.com, www.borgenergy.com

SULAKSHANA CIRCUITS LTD

Durga Rao, MD

Established in 1988, Sulakshana Circuits manufactures double-sided and multilayer PCBs of high quality and in medium to high volumes. Headquartered in Hyderabad, its plant is spread over 5 acres, where high-quality imported materials, modern manufacturing processes, experienced managers, skilled personnel and imported capital equipment combine to produce the highest quality boards at competitive costs.

With four different bare board testing machines (BBT), the company can cost-effectively test all boards before shipment.

Sulakshana Circuits grew 22 per cent in FY 2012-13 and is looking to double its pace of growth in 2014-15 with more new equipment. It was recognised as a 'zero defect supplier' by BEL for FY 2011-12.

Automotive, instrumentation and controls, railways, telecommunications, medical, aerospace, defence, consumer and contract electronics

Contact details: Plot No. 36 & 37, Anrich Indl. Estate I.D.A. Bollaram, Hyderabad - 500 032, Telangana Ph: 040-3242 0340, sivakumar@sclpcb.com, www.sclpcb.com

AVNI ENERGY SOLUTIONS PVT LTD

G Gururaja, founder director

Avni is focused on manufacturing LED lights, fixtures and drivers. Its customer service, backed by strong in-house R&D, has made it one of the top manufacturers of LED lights in India

Started in 2009 by G Gururaja, Avni Energy Solutions Pvt Ltd is now one of the leading LED manufacturers in India. It is an ISO: 9001-2008 certified company with a major facility in Bengaluru and a satellite plant in Raipur, Chhattisgarh.

Avni is focused on manufacturing LED lights, fixtures and drivers. Its customer service, backed by strong in-house R&D, has made it one of the top manufacturers of LED lights in India. It has a highly qualified and skilled technical team.

Avni's LED lights are designed for residential, commercial, outdoor and other infrastructure lighting applications. Its products are affordable and compliant to stringent standards. Avni offers intelligent solutions that use renewable energy such as wind and solar.

Its credentials are backed by certification from NABL accredited labs like CPRI, ETDC-Bengaluru and ERTL-Kolkata. For professional light simulations, Avni uses specialised software such as Relux or DIALux.

G Gururaja, the founder and director of Avni, is also the vice president of LEDMA (LED Manufacturers Association) and a panel member of BEE, working towards standardisation in LED lights in India.

Strengths
- Outstanding application engineering support with the capability of offering customised solutions to clients
- High volume manufacturing capabilities
- Affordable products of the highest quality
- A passion for continuous innovation to drive the adoption of energy saving technologies

Key achievements
- It is ranked No 3 among Indian LED product manufacturers, having crossed Rs 500 million in turnover within 4 years
- Winner of ET Now's 'Leader of Tomorrow-2013' award
- Avni has supplied more than 20,000 LED streetlight installations to over 20 municipal corporations across India
- It is a technology partner of TERI and is one of the preferred vendors for its LABL campaign
- Preferred partner for leading LED and IC manufacturers for LED technology development
- Has supplied more than 200,000 solar lanterns to rural areas; sold over 100,000 downlights

Contact details: 39/5, 8th Cross, Govardhan Garden, Opp Delhi Public School, JC Industrial Area, Near Kanakapura Road Metro, Bengaluru 560062, 080-26860337, gururaja@avnienergy.com, www.avnienergy.com

NURA LED

Jahir Hussain,
MD

Founded in 2008, Nura LED is a major provider of LED lighting solutions. Leveraging the group's five decades of experience, Nura has successfully become a leading provider of LED lighting solutions for an entire range of applications.

Nura's constant focus on LEDs has motivated its team to develop green lighting solutions that reduce operating costs and improve productivity.

As the leading innovator in the LED lighting industry, Nura has a proven track record of helping companies realise the potential business benefits of LED lighting solutions.

Nura has also played a pioneering role in the areas of quality of service and light output by combining the most advanced digital control, thermal management and lighting fixture designs with international quality standards and performance.

Nura provides the most versatile, best-in-class and cost-effective LED lighting solutions for various applications.

Contact details: No. 432, Nanjundapuram Road, Sri Raam Plaza, 1st Floor, Ramanathapuram, Coimbatore - 641045, nuraenterprises@gmail.com, Ph: 9600738595

BOSCH SECURITY SYSTEMS

Sudhir Tiku,
head–South Asia

Bosch Security Systems in India has offered state-of-the-art safety, security and communication solutions to various prominent projects like metro rail projects, surveillance solutions to numerous airports, city/market, etc

Bosch Security Systems is committed to securing people and premises, wherever lives and property are at risk. Its customers choose it for the innovativeness and efficiency of its products, and for the Bosch reliability and high quality of workmanship. The company also offers a wide variety of products, all backed by a worldwide presence.

Products offered

Video surveillance systems: The innovative video surveillance solutions from Bosch are a perfect fit for a vast array of applications: from multi-site, multi-user configurations for large commercial and governmental applications, to a lighter range for small businesses. Bosch provides comprehensive solutions to capture, analyse, compress, transmit, view, store, search, export and integrate videos and images.

Access control systems: Bosch provides first-class access control software that integrates the leading-edge security technologies with innovative networking capabilities, offering customers full featured security solutions serving applications of any size. Further, Bosch can also provide building integration systems (BIS). With BIS, efficient building management is made simpler – combining different building management functions on one platform, and providing simple responses to difficult problems.

Fire alarm and evacuation systems: Bosch has more than 80 years of experience in developing fire alarm systems, and now has a product portfolio to suit most applications. To ensure optimal protection, an efficient evacuation system is also necessary. Bosch offers a range of highly robust EVAC (Emergency Voice Alarm Communication) systems for sites of all sizes, including Praesideo – the world's first fully digital public address and emergency sound system.

Public address and conference systems: With over 50 years of experience in the public address and emergency sound field, Bosch is able to offer innovative product solutions for any public address and voice evacuation requirement. These products offer excellent value for money, extensive yet highly user-friendly functionality, and stylish, contemporary design.

From intimate informal meetings, to major multi-lingual conferences, Bosch Conference Systems facilitate effective interaction and communication in a way that suits everyone.

Pro-sound and critical communications systems: Bosch Communications Systems also owns two pro-sound brands—Electro-Voice (EV) and DYNACORD, as well as two critical communications systems brands, TELEX and RTS. EV and DYNACORD are universally recognised as innovators in the professional audio industry. Focused on critical communications systems, TELEX and RTS have the longest history and most complete product portfolio in this field.

Bosch Security Systems in India has offered state-of-the-art safety, security and communication solutions to various prominent projects like metro rail projects, surveillance solutions to numerous airports, city/market surveillance and traffic management solutions in major metros, refineries, industrial complexes, residential apartments, etc.
Contact details: P B No - 3000, Hosur Road, Adugodi, Bangalore - 560 030, India, Ph: 080-67528143, santosh.kumar@in.bosch.com, www.boschsecurity.co.in

UNIQ POWER SOLUTIONS

R Srinivasan, proprietor

Established in 2010, Uniq Power Solutions is an importer of reputed sealed maintenance-free, tubular, Ni-Cd, Ni-Mh and lithium cells and packs in South India. It offers a comprehensive range of batteries for diverse applications. Its capabilities enable it to cater to the various needs of the UPS system, security, telecom, medical equipment, solar power system, elevator and emergency light segments. Other devices that its imported batteries cater to are emergency lights, weighing scales, attendance registers, instruments, vending machines, etc.

Contact details: 240, 8th Main Road, Byrasandra Extn, 1st Block (E), Jayanagar, Bengaluru, 560011; Ph: 080 - 26645806/9845135538; info@uniqpowersolutions.in

HK WENTWORTH (INDIA) PVT LTD (ELECTROLUBE)

Padmanabha Shakthivelu,
Electrolube Sales Manager - India

Electrolube offers a vast array of products including thermal management materials, conformal coatings, encapsulation resins, electronic cleaning solutions and general maintenance products

Electrolube's roots can be traced back to 1941, when Henry Kingsbury formed Kingsbury Components to manufacture volume controls. It was then that he formulated a special oil, which enhanced the electrical performance and lifetime of the contact surface, in addition to reducing friction of moving parts. This breakthrough was the catalyst for a whole range of contact lubricants, forming the basis of the company now known as Electrolube.

Products offered

As a global specialist in formulated chemical products for the electronics industry, Electrolube offers a vast array of products including thermal management materials, conformal coatings, encapsulation resins, electronic cleaning solutions and general maintenance products.

Its wide product range enables it to supply to leading manufacturers of electronic and industrial devices, offering the 'complete solution' at all levels of production. With a strong emphasis on both research and collaboration, Electrolube is constantly developing environment friendly solutions for its customers, both old and new.

With a presence in 55 countries and expanding, its ethos in innovation and service remains strong. The firm's vision is: "To strive to exceed our customers' expectations with innovative new products and the highest possible levels of customer service." All its product ranges are manufactured at its factories in the UK and China to ISO 9001, ISO 14001 and OHSA 18001 certification.

Customer service

Since it is present in over 55 countries, its disciplined network of subsidiaries and distributors can offer all its customers genuine security of scale. This robust supply chain means that when the unexpected happens, Electolube is still capable of delivering a truly bespoke service anywhere in the world.

Its unrivalled, very personal customer service also extends to every corner of the world. Customers can rely on its fully trained staff to be knowledgeable and up to speed with the latest technological developments. Having a named single point of contact helps ensure it solves problems before they become issues.

Sectors it caters to

• Automotive • Military • Aerospace • Transport • Marine • Telecommunications • Medical • Consumer electronics • Industrial electronics • Traction • Utilities • Education • Service, repair and maintenance

Contact details: *Private office 110, DBS House, #26, Cunningham Road, Bengaluru 560052, Ph: 080-40407110, padmanabha.s@hkw.co.in, www.electrolube.com*

PROGRESSIVE TECHNOLOGIES

T Pramod Kumar, CEO

Progressive Technologies manufactures LED lighting for all types of requirements. Its products include home lighting equipment for remote areas with no power or insufficient power for day to day requirements. To offer complete solutions under one roof, the company has started solar panel (module) manufacturing. It also undertakes EPC contracting of solar power projects, roof-top solar power installations for the domestic, commercial and industrial sectors, and has now entered the field of MW scale solar power plant erection and commissioning.

The company is ISO 9000 certified, CRISIL and SMERA rated, and NSIC registered. Its products are manufactured in a clean and air-conditioned environment, and go through stringent test procedures to ensure consistent good quality.

Progressive Technologies is a proprietary company managed by T Pramod Kumar, who is qualified in electrical and electronics engineering and has 33 years' experience in the design, development and manufacture of power conservation equipment.

Contact details: *MIG-9, NFC Main Road, HB Colony Meerpet, Moulai-Ali, Hyderabad 500051, Andhra Pradesh, Ph: 09395524039*

EVERLIGHT ELECTRONICS CO LTD

Giri Prasad G V, country manager

Everlight is one of the few global optoelectronics suppliers to have complete control over its production facilities. This in-house capability enables it to have full manufacturing and quality control

Everlight Electronics Co Ltd was founded in 1983 in Taipei, Taiwan. Playing a critical role in the formation of the global LED industry, the company has rapidly grown to become a leading supplier due to its dedication to certification, R&D, production, quality, marketing and global customer service.

Today, Everlight is a global company with over 5,600 employees, and has its headquarters in Taipei, Taiwan. Everlight sales offices are located around the world (in China, Hong Kong, Japan, Korea, Singapore, Malaysia, India, Germany, Sweden and USA) to provide immediate service and prompt delivery to its customers. The company's manufacturing facilities are located in China (Suzhou, Guangdong) and Taiwan (Tu-Cheng, Yuan-Li).

Everlight Electronics' excellent performance proves that it can overcome hurdles. In addition to having a diverse product portfolio consisting of high power LEDs, lamps, SMD LEDs, lighting modules, digital displays, optocouplers and infrared components, Everlight Elecronics' ability to integrate both downstream and upstream development enables it to lead the market.

Everlight is one of the few global optoelectronics suppliers to have complete control over its production facilities. This in-house capability enables Everlight to have full manufacturing and quality control, ensuring a quick reaction to market shifts, with short lead time responses.

Everlight's centralised R&D teams are continually improving and developing products to meet market demands. As a 'complete service opto-components provider', the company understands—and delivers—what today's designers need, including comprehensive engineering specs, safety compliance and electrical specs, as well as RoHS-compliant packages.

With its advantages in manufacturing and cost control, Everlight will continue enhancing its competitiveness by diversifying its product portfolio and securing its intellectual properties through patent procurement. To meet the rising demands of the global optoelectronics market, Everlight will continue to strengthen its R&D base in order to achieve major technological advancements in its next generation LED products–to improve functionality and reliability in tomorrow's applications.

Contact details: *Suite No 3, 2nd Floor, Alpha Block, Sigma Soft Tech Park, Varthur Kodi, Whitefield Main Road, Bengaluru 523860, salesindia@everlight.com, www.everlight.com*

NICHANI ELECTRONICS

Deepak Nichani, MD

Established in 1993, Nichani Electronics has specialised in electronic components distribution across the south Indian states. It has a great team that selects quality components, and manages administration and logistics, enabling the firm to strike the right balance between quality, price and service.

The company's strength lies in passive components, especially capacitors – electrolytic, screw terminal, metallised polypropylene, ceramic and tantalum. It also specialises in electronic hardware like instrument-cooling fans, terminal blocks; rocker, toggle and micro switches; fuse holders and fuses.

It is focused on supplying 105 degree 5,000-10,000 hr aluminium electrolytic capacitors for LED driver and power supply applications. It offers top quality and high reliability capacitors.

The company's aim is to satisfy its customers on all counts. It has a customer service team that is quick and responsive to enquiries and proactive when it comes to requirements. The firm believes that action taken based on innovation leads to excellence–by pursuing excellence, success will definitely follow.

Contact details: *Ph: 044-28544226/ 42131889/ 28586450, nichanielectronics@ rediffmail.com, www.igbt.in*

ELEKTRONIKA SALES PVT LTD

Sunil Hasija, MD

Elektronika Sales is planning to create a footprint in the global market through the acquisition of components distributors overseas

Established in 1979, Elektronika Sales Pvt Ltd is one of the largest electronic components distributors in India. The company caters to OEMs across India in the automotive, energy meter, power, lighting and industrial segments.

Elektronika has its sales offices across India and caters to over 1200 customers. It supports the industry through its skilled and qualified sales and marketing team, and field application engineers. In addition to this, it provides technical solutions support through its design house, which specialises in power, embedded and RF/Bluetooth modules.

Services offered
- Active, passive and electro-mechanical components under one roof
- Specialises in kitting and consolidation
- VMI programme for key OEMs
- Design services for RF, embedded and power modules
- Technical support with reference designs
- INR/USD billing

Major milestones
Elektronika Sales is the proud recipient of various awards from suppliers and customers.
- VISHAY - Best Distributor Award 2012
- NXP - Silver Award Winner 2011
- VISHAY - Best Distributor Award 2011 and 2012
- VARROC - Best Supplier Award 2011
- KAYNES / VINYAS – Best Supplier Award

Solutions from Elektronika's design house
- RF module based on NXP Zigbee for an energy meter (TUV certified)
- A Bluetooth module based on a Toshiba chip
- LED driver solution from 5W ~ 140 W based on a power integration chip
- Solutions for battery chargers, RO adaptors, CCTV cameras, modems, set-top boxes and power supplies

Vision
Elektronika's vision is to be the leading technology and solutions provider. Its road map is to be a Rs 5000 million company in the next five years. Elektronika Sales is planning to create a footprint in the global market by acquiring a network of components distributors overseas.

Contact details: 16, Narasingapuram Street, Mount Road, Chennai 600002, Ph: 044-28587165/7765, info@electronika-sales.com, www.electronikasales.com

PACIFIC ELECTRONICS PVT LTD

P V Raghava Rao, MD

Pacific Electronics Pvt Ltd was established in 1978 by the late P Hanumantha Rao, one of the top most technocrats in India and recipient of the Udyog Patra award in 1987. After his demise, his wife, the late Annapurna, took on the responsibility of the company. At present, P V Raghava Rao is the MD of the firm.

Known for manufacturing the best electronic test and measuring equipment in south India, Pacific Electronics is also the only company that manufactures CROs in India. Its other products include the Aquila lab trainer kits. With 34 years' experience, Pacific Electronics is launching a new product, Power Saver PPS-216, which saves electricity. This is connected between the AC and a ceiling fan, which are operated alternatively and automatically, reducing the electricity consumption. This product consumes maximum power of just 1mV. Its output is controlled through a contactor of rating 32 amperes, while time control is done through a microcontroller.

Pacific is also introducing a new mini lab (six-in-one), which consists of a CRO, power supply, function generator, voltmeter, ammeter and frequency meter.

Contact details: Plot No 22, Ravi Co Op Society, Opposite SBI Trimulgherry Branch, Trimulgherry, Secunderabad 500015, Ph: 040-27791139/9848058904, audiogen@usa.net, www.pacificelectronics.in

LWI ELECTRONICS INC

Dinesh Singh Samyal,
CEO

The adoption of an innovative and practical approach to independent distribution enables LWI to deliver quality consistently

LWI Electronics Inc is an ISO 9001:2008 certified organisation that has grown to become the preferred independent distributor of choice for numerous customers. Since 1999, LWI has built a reputation for itself as a highly reliable provider of semiconductors and allied services. The company is committed to customer satisfaction, which is reflected in its business activities, like interacting and negotiating with manufacturers, maintaining inventory or providing prompt customer service.

The adoption of an innovative and practical approach to independent distribution enables LWI to deliver quality consistently. The company was recently ranked as one of the world's Top 50 independent stocking distributers by ISCP, Canada. It has acquired ERAI certification, USA, and ESC membership, India.

Inventory of line items

With a growing middle-class of nearly 400 million people, India's electronic equipment consumption, which was estimated at around US$ 210 billion in 2013, is expected to reach US$ 380 billion by 2015. Hence, LWI stocks a vast inventory of line items, which makes it the No 1 stocking distributor in India. Today, LWI stocks over 150,000 line items of new and obsolete semiconductor components.

Products offered

LWI provides total semiconductor, networking and connectivity solutions for industry and the defence segment. It distributes military and industrial grade parts and products like integrated circuits, capacitors, resistors, discrete devices, linear logic, digital signal processors, DRAM, SRAM, cache memory, connectors, LAN products, and much more. It expects to double its line items by 2015.

State-of-the-art facility

With over a dozen franchisees all over the globe, LWI has got a state-of-the-art 24 x 7 updated stock search facility with in-house component test labs that enable it to improve on quality. It also has a testing facility for obsolete parts.

LWI has invested in the latest technical equipment and quality manpower that deliver results, year after year. Its experienced technical team is well versed in providing turnkey solutions in electronics to its customers.

Contact details: *62, 14th Cross, 1st Main, I Block, R T Nagar, Bengaluru 560032, Ph: 080-23530578, sales@livewireinfo.com, www.livewireinfo.com*

AMPTRONICS SYSTEMS PVT LTD

Mitesh Patel, CEO

Established in 2009, Amptronics Systems Pvt Ltd (ASPL) is a fast growing, leading solutions provider catering to the oil and gas, water recycling, power generation and heavy equipment industries. ASPL's solutions take distributed asset monitoring and assessment to a new level of accuracy and timeliness, with a track record of transforming data into useful information that helps customers improve operational excellence.

The company's strength lies in its R&D team, which is constantly working on providing innovative products and solutions. ASPL is a design house providing concept-to-end solutions in the product design and production space.

ASPL has earned the reputation of providing high-performing products by

controlling all aspects of the development process including design, manufacturing and testing. Quality assurance has always been the watchword of its manufacturing process.

The company aims to be the leader in designing, manufacturing and providing global data communication products, services and solutions that expand the connected world.

The sectors it caters to are: energy, automotive, transportation and fleet management, industrial and infrastructure, networking, sales and payment, security, field service, healthcare and consumer.

Contact details: *No. 6-3-905/B2, Jaffer Ali Bagh, Somajiguda, Hyderabad 500082, Ph: 040-40077096/23320240, info@amptronics.com, www.amptronics.com*

DIGITAL CIRCUITS PVT LTD

Subhash Goyal, director

Digital Circuits Pvt Ltd (DCPL), a leading electronics manufacturing services (EMS) provider, is headquartered in Bengaluru. DCPL has been an electronics hardware technology partner to the automobile, telecom, consumer electronics, medical electronics, instrumentation and electronic metering industries for about three decades.

Compliance to standards

It is a UL DQS, ISO/TS 16949:2009 and ISO 9001:2008 certified company with a clear focus on quality, cost and customer delight. It also follows WCM practices like 5S, Kaizen, Poka Yoke, and JIT manufacturing.

State-of-the-art infrastructure

Its facility includes state-of-the-art infrastructure for SMD, pick and place, chip shooters, automatic solder paste printers, reflow ovens, wave soldering, AOI, BGA/micro BGA work stations and ICTs.

DCPL's capabilities and core strengths include SMT, through-hole and mixed technology, plastic injection moulding, sheet metal fabrication, wiring harness, box build, and testing, including environmental tests.

Hence, it offers all electronics manufacturing solutions under one roof, starting from the design of the product.

Manufacturing facilities

It has manufacturing units in Bengaluru, Baddi (Himachal), Rudrapur (Uttaranchal), Pune, Mysore, Delhi (Manesar) and Chennai, which reach out to more than 100 customers spread all over India. DCPL exports to countries like USA, Europe and Israel.

Contact details: *No. 60, Doddakala Sandra, Kanakapura Road, Bengaluru 560062; Ph: 91-80-22560995, 9845026472; subhashgoyal@digitalcircuits.in*

RECOM ASIA PTE LTD

Karsten Bier, CEO

RECOM, a leading supplier of converters in the power range of 0.25 W to 60 W, offers a portfolio of 22,000 DC/DC converters, AC/DC converters, switching regulators and LED drivers. These converters are available from the world's leading distributors. The high performance products are engineered in Europe and offer high isolation, high ambient temperatures and high efficiency combined with a low cost of ownership.

"At RECOM, we have a very simple quality objective: Zero defects," says Karsten Bier, CEO of the RECOM Group. "We make no compromises and use the best parts available in the market," he continues.

All prototypes are HALT tested in the company's environmental laboratory prior to release. Once a product is in the market, samples are re-tested on a regular

basis to ensure 'Total Quality'. Some of the products have a tested 'design lifetime' of > 70,000 hours, which is why the company can easily offer a product warranty of up to 5 years.

The company has made significant investments in new products during the past months and is set to grow faster than the industry average.

The company's growth is driven by its well-established practice of using certified converter modules rather than discrete solutions. "Time to market is what counts for most of our customers," says Bier, "even if the BOM is a little higher."

Contact details: *60 Kaki Bukit Place, 05-01 Eunos Techpark, Singapore 415979; Ph: +65 6276-8795; Fax: +65 6273-1477; enquiry@recomasia.com; www.recomasia.com*

SOUTHERN BATTERIES PVT LTD

R Sreenivasan, MD

Southern Batteries' Hi-Power brand of batteries is suitable for various applications like lead acid tubular batteries, valve regulated lead acid, etc

Inspired by the vision of its founder and driven by the resolve to bridge the demand-supply gap, Southern Batteries Pvt Ltd was founded in 1980. Within a short span of time, the company made its mark as a quality player in the secondary power storage industry. Today, it is one of India's leading manufacturers of tubular batteries.

Southern Batteries has state-of-the-art manufacturing units and the entire process is done in-house, right from the plates to final assembled batteries and charging. Its Hi-Power brand of batteries is suitable for various applications like lead acid tubular batteries, valve regulated lead acid (VRLA) batteries, flat plate batteries, traction batteries and automotive batteries.

Hi-Power products are approved and certified by the Indian Railways. They are sourced by all major solar players in India for solar photovoltaic applications. Hi-Power is also used by almost all UPS system and inverter manufacturers, state electricity boards, the Nuclear Power Corporation, banks, government organisations and educational institutions.

Southern Batteries has forayed into the retail market and is marketing its products through a national network of branches and channel partners. To ensure total customer satisfaction, the company has an exclusive 24/7 customer service division managed by well-trained personnel.

As a firm with a keen sense of corporate responsibility, it actively participates in solar power projects and works closely with manufacturers of electric vehicles. Southern Batteries seeks to reinforce its leadership by expanding its horizons with a focus on innovation, quality and sustainability.

Southern Batteries is certified for ISO-9001:2008, ISO-14001:2004 and BS-OHSAS-18001:2007.

Contact details: *No. 328, Bommasandra Jigani Link Road, Anekal Taluk, Bengaluru 562106, Ph: 080-22010000 to 22010099, enquiries@southernbatteries.com, www.southernbatteries.com*

MOUSER ELECTRONICS

Mark Burr-Lonnon, vice president, APAC business

Mouser Electronics, a wholly owned subsidiary of Berkshire Hathaway Inc, is one of the industry's fastest growing global catalogue and online semiconductor and electronic components distributors. It has its headquarters in the US, and its Asia main branch is located in Hong Kong. Dedicated to supplying design engineers and buyers with leading edge technologies backed by unsurpassed customer service, Mouser is redefining customer focused distribution. The company works in close partnership with all its suppliers to provide the fastest access to the industry's latest components, giving design engineers a crucial technological edge and the speed-to-market advantage. Having the latest technology to develop cost efficient prototypes limits costly re-designs, manufacturing delays or even the termination of a project.

Mouser.com searches nearly 8.9 million products to locate over 3 million part numbers, shipped from its state-of-the-art 45,700 sq m (492,000 sq ft) warehouse. It offers its services in 17 languages and accepts payment in 17 international currencies. Mouser.com clearly identifies parts that are not recommended for new design (NRND) and features helpful tools like the 'BOM tool' and 'Mouser search accelerator'. Its broad-based line card consists of millions of advanced components across the board with each vertical completely covered. In addition to its online version, Mouser's catalogue is printed twice a year, removing any obsolete parts. Customers are always confident of designing with the most advanced electronics available.

The company excels in providing service and support to customers worldwide. It operates out of multiple locations around the globe. Mouser takes orders without the requirement for a minimum quantity, offers rapid delivery, live chat in 7 languages, technical and product support from knowledgeable teams, and more.

Contact details: *Ph: +91 80 41148091/92; india@mouser.com*

STATIC SYSTEMS PVT LTD

G Chandra Shekhar,
director

Static Systems helps electronics companies to set up ESD protected areas by providing an ESD flooring, supplying conductive furniture, etc

Static Systems Pvt Ltd is an ISO 9001:2008 certified company that was established in 2002 to provide 'total solutions for the electronics industry.' It imports ESD safe products, soldering equipment, tools and machinery required in PCB assembly.

Services

Static Systems helps electronics companies to set up ESD protected areas by providing an ESD flooring, supplying conductive furniture, etc. It also offers ESD consumables like ESD aprons, ESD shoes, ESD bins, ESD covers, etc. All the ESD products conform to ANSI /ESD-EOS-S-20.20-1999 standards.

Foreign tie-ups

Werk Sitz, Germany: Specialises in ergonomic swivel ESD work chairs, which take care of users' health and increase performance.

MicroCare, USA: This firm is considered a leader in precision cleaning electronics for the manufacturing process, rework or repair.

Xuron Corp, USA: This is one of the world's leading manufacturers of ergonomic electronics-grade hand tools for the electronics, aerospace, plastic moulding, hobby, craft and jewelry industries. Xuron sets the standard for precision, tool durability and ease of use.

Musashi Engineering Inc, Japan: For manual dispensers, robotic dispensers and soldering robots, Musashi guarantees high performance, high quality, cost reduction, high productivity and compact sizes.

OZU Corporation, Singapore: Bemcot wipers and related products serve cleanroom applications through a unique synergy of characteristics intrinsic to the products' composition, structure and purity.

Scienscope, USA: This is the leading NDT (non-destructive testing) supplier of microscopes, industrial metrology systems, video measurement systems and high resolution cabinet X-ray systems.

Contact details: *Sri Krupa, No. 925/3, I Floor, 3rd Main, 3rd Cross, 2nd Stage, 'D' Block, Rajajinagar, Bengaluru 560010, Ph: 080-41285756/23133597, 9880203636, staticsystemspvtltd@gmail.com, www.staticsystem.com*

TEMCO ELECTRICALS & ELECTRONIC INDUSTRIES

Sreejesh M P, CEO

Temco Electricals & Electronic Industries is an ISO 9001-2008 certified company that specialises in solar power solutions and LED lighting. Its vision is to be recognised as the premium solutions provider in the solar power and LED lighting space.

Cost effective solutions

Temco specialises in solving energy related problems by offering cost effective and reliable power solutions. Its in-house R&D team ensures that its products have the latest and the most reliable technologies implemented so as to achieve complete customer satisfaction.

All its systems come with an inbuilt isolation transformer, which makes them extremely rugged to face critical environmental conditions. Temco is one of the few solar off-grid inverter manufacturers to be accredited by IEC-61683 for efficiency tests and IEC-60068 for environmental tests.

Product range

It manufactures solar PCUs, MPPT charge controllers, PWM charge controllers, solar hybrid UPS systems, DSP sine wave UPS systems, solar LED street lights, LED bulbs, LED tube lights, etc.

Contact details: *No. 72, 1st Floor, Mottappa Compound, RMV 2nd Stage, Devinagar, Sanjay Nagar P.O., Bengaluru 560094, Ph: 07795552777/080-32010525, temcoblr@gmail.com, www.temco.co.in*

SAKTHI ACCUMULATORS PVT LTD (FORMERLY SAKTHI ELECTRONICS)

R KANDASAMY, MD

Sakthi is looking forward to expand its market share for tubular batteries, particularly in the solar and automotive battery segments

Established in 1992, Sakthi Electronics is committed to consistently manufacture and supply products of outstanding quality and reliability along with efficient service support, to the utmost satisfaction of its customers. It works towards continuous improvement in every functional area.

Focus on quality

The company has earned the full commitment and participation of all its employees, which encourages innovation and total quality. Its mission is 'total customer satisfaction'. Its products are manufactured with strict quality control in all the processes of manufacture, in the components used, and in employing modern engineering and manufacturing practices. Manufactured products are based on state-of–the-art technologies developed at its in-house R&D centre, while also incorporating technology and knowhow from external resources.

Product range

- Tubular batteries • Automotive batteries
- Traction batteries
- Batteries for solar applications
- Stationary tubular cells
- Aircraft starting batteries
- Sealed maintenance-free batteries
- Lead acid batteries

Compliance to standards

- Its tubular batteries are made to conform to IS:1651-1991 and IS:13369-1992 standards.
- Its tubular batteries for solar applications are made to conform to MNES specs, and are tested by ERC and TVM.
- Its automotive batteries are made to conform to IS:7372-1995 and IS:6304-1992 standards.
- Its traction batteries are made to conform to IS: 5154-1980 standards.

Future plans

Sakthi is looking forward to expand its market share for tubular batteries, particularly in the solar and automotive battery segments. It also plans to establish a separate R&D department to develop various types of batteries. The company will be taking up the development and supply of batteries (type MT and WT) to the defence sector.

The company recently changed its name to Sakthi Accumulators Pvt Ltd.

Contact details: 231, 1st Main, 3rd Cross, KSRTC Layout, J P Nagar, 2nd Phase, Bengaluru 560078, Ph: 080-26588251, Office@ sakthipower.com, www.sakthipower.com

BULJIN ELMEC PVT LTD

V Murali, MD

Buljin Elmec Pvt Ltd was incorporated in 1986 to manufacture PCBs. It later diversified into flexible PCBs to cater to the needs of the Indian auto sector. In 2008, it began manufacturing metal core PCBs (MC PCBs) for LED light applications. It currently provides MC PCBs to both the Indian and international markets. Quality and on-time delivery are the firm's guiding principles.

Services offered

- PCB design
- V-grooving
- Tool punching
- Reflow oven for soldering SMD LEDs

Products offered

It also supplies the full range of LED lights and fixtures for LED bulbs (1-12 W), downlights, tube lights, street lights, flood lights, high bay lights, para-square ceiling lights, etc.

Contact details: 2/10, Krishna Industrial Estate, Mettukuppam, Vanagaram, Chennai 600095, Tamil Nadu, Ph: 044-24764774/24764657, Mobile No: 9841698491/ 9176659966 / 9841698490, buljin2000@yahoo.com, info@buljin.com, www.buljinledlights.com

OMNISCIENT ASSOCIATES

Mahadev Prasad,
senior product manage

Omniscient Associates is a leading Indian distribution company with a product portfolio covering speciality tapes, adhesives and epoxies, interconnect solutions, ESD products, electronics grade coatings and chemicals.

Omniscient's product portfolio holds the leading position in a variety of segments like electronics, electrical and industrial. It caters to markets including general industrial, aerospace, railways, defence and space, automation, telecom and EMS manufacturers.

The company operates as a dynamic and proactive organisation, responding instantly to market demands. As a result, it is able to supply the highest quality components at the best possible price. This is combined with an outstanding level of customer service.

Customer service

Omniscient Associates has always endeavoured to offer its customers the very best value for money. Its product range, quality and price are the key factors that ensure customer satisfaction. Omniscient's customers expect and receive unbeatable value and appreciate the company's commitment. Its engineers support customers by providing the design tips that help them create products with ease, using Omniscient products.

Employees: An asset

Omniscient Associates has consistently achieved excellent results in a highly competitive market place. The company appreciates its people for their dedication and initiative in delivering a first class level of customer service. Their enthusiasm has been an inspiration, forming the bedrock on which Omniscient has grown.

Contact details: *Anvin House, #782, 3rd Main, 1st Cross, 5th Phase, BEML Layout, Rajarajeshwarinagar, Bengaluru 560098 Ph. 080-28611874, prasad@omniscientassociates.com*

KWALITY PHOTONICS PVT LTD

Vijay Kumar Gupta,
CEO and MD

Kwality Photonics Pvt Ltd is India's largest producer of LEDs, LED displays and opto electronic products. Kwality is the first company to have successfully established LED production in India and commands the highest share of the domestic market.

The company was established in 1966 under the guidance of its CEO, Vijay Kumar Gupta, who is a noted scientist engineer and expert in LEDs, having been associated with LED manufacturing technology for over 30 years.

Currently employing more than 250 workers, the company started off with the manufacture of lamps and lamp components. Production of LED displays began in 1987 after successful indigenous R&D.

Kwality is a technology and quality leader due to its ability to deliver state-of-art products. It is ISO 9001: 2008 certified and consistently achieves near 100 per cent yields, which has enabled the firm to offer over 600 types of LEDs, LED displays and LEDs for lighting at the most competitive prices, ensuring the best value for money for its customers.

Products offered

The optoelectronic product range includes LEDs, LED displays, LED lamps, IREDs (infrared emitters), photo transistors, photo detectors, flasher LEDs, hi-voltage LEDs, low battery indicator LEDs, hi-flux LEDs, LED clusters and filament lamp retrofits.

The Kwality Photonics PolyWa LED series delivers the highest lumen efficiencies in the globe. PolyWa 3 W 280 Lm, 1 W 130 Lm, half-watt 5630 65 Lm, 2835 W 24 Lm, and 3528 W 8 Lm LEDs and COB

Kwality Photonics has recently launched uniLED, India's first and only LED automotive fitment bulb designed to be used in every type of OEM automotive light across all categories of vehicles

LEDs, come with single bin uniformity in brightness and colour value, all at globally unbeatable prices.

The PolyWa 5630 LEDs are ideal for retrofit applications, aimed at substituting 60 W bulbs and 36 W CFLs. These power LEDs have a copper slug placed right under the LED chip to provide the shortest path for the heat to dissipate, and have the lowest thermal resistance in their class as they are manually solderable. The 8 Lm PLCC3528 LEDs are the firm's latest tube light offerings and come with aggressive pricing.

Kwality Photonics has recently launched uniLED, India's first and only LED automotive fitment bulb designed to be used in every type of OEM automo-

tive light across all categories of vehicles (2,3 & 4 wheeler) and across all models (past and present) within the automotive industry. Lasting upto 50,000 hours, these bulbs are the only LED bulbs that have optimised light beam focusing properties to meet the requirements of long lasting standard indicator and tail light fitments. The series offers features such as rapid brightening, energy efficiency, being environment friendly, and water and heat resistance.

Contact details: 29, Electronics Complex, Kushaiguda, Hyderabad 500062, Ph: 040-27123555/ 09391017046/09000081171, kwality@kwalityindia.com, www.kwalityindia.com

CIPSA TEC INDIA PVT LTD

Alok Garg, MD

CIPSA TEC always aims to build long-term relationships, and the result is that it has customers with whom it has been associated for more than 10 years

CIPSA TEC India Pvt Ltd is one of the largest PCB manufacturers in India. It is a joint venture between CIPSA, Spain; TECNOMEC, Italy and its Indian promoters. These two partners have remained pillars of strength in the company's growth, particularly in technology and process improvements.

CIPSA TEC's current manufacturing facility is a fairly new plant that was set up by Siemens in Germany. In 2008, this plant was relocated to India near Bengaluru (Tumkur).

Currently, the plant's capacity is 16,000 sq m and it is spread across an area of 36,000 sq m. This year, the company plans to expand its manufacturing capacity up to 20,000 sq m. Today, CIPSA exports 20 per cent of its production to Europe.

The company manufactures PCBs ranging from DS up to 8 layers (ROHS as well as non-ROHS) and also metal clad PCBs. It supports prototype development (delivering within 4-7 days), quick turn-around production (8-10 days) as well as production within normal lead times of 3-4 weeks. It can deliver quantities ranging from 1-1 million PCBs.

The company's core strength lies in its excellent quality and consistent delivery at competitive prices. It is open to work with customers on R&D and PCB related issues

and better penalisation, which results in improved productivity.

The company has an excellent communications system which ensures faster quotations, order acknowledgements, pending order status (weekly) reports, despatch updating, etc. Its QMS conforms to ISO:9001:2008 and TS-16949 (TUV certified) standards; and its products conform to UL (USA and Canada) and the Indian government's standards for telecommunications and defence..

It always aims to build long-term relationships, and the result is that it has customers with whom it has been associated for more than 10 years. It caters to most of the segments including telecom, energy, power, energy meters (single and three phase), UPS systems, inverters, automotive (clusters), ECUs, thermo controls, sensors, lighting, audio, security, industrial and medical—both EMS and for OEMs.

The company is aiming at a growth rate of more than 20 per cent this year.

Contact details: Plot Nos 7 and 8, Hirehalli Industrial Area, Tumkur 572168, Ajaish Sehgal, senior manager-marketing, Ph: 09940018012/ 09790810512, ajaish@cipsatec.com, www.cipsatec.com

NEW PRODUCTS ▶▶

TEST & MEASUREMENT

Oscilloscope series

In May 2014, **Agilent Technologies** introduced two new high-performance portable oscilloscope series that have incorporated next-generation oscilloscope technology. The Infiniium S Series sets a new

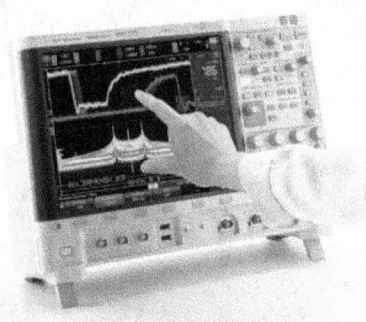

standard in signal integrity for bandwidths up to 8 GHz, while the InfiniiVision 6000 X-Series sets a new standard in price and performance for bandwidths up to 6 GHz. The Infiniium S-Series has the world's fastest 10-bit ADC. It features a comprehensive application-specific measurement software, fuelled by a powerful motherboard with 8 GB RAM, ensuring the scope stays responsive in all operating modes. The InfiniiVision 6000 X-Series includes 2- and 4-channel DSO models, and MSO models with 16 digital channels and with bandwidths from 1 GHz to 6 GHz at a 20-GS/s sample rate.

For further details: Ph: 011-46237100, tm_india@agilent.com, www.home.agilent.com

Digital oscilloscopes

In May 2014, **Rohde & Schwarz** introduced its RTE digital oscilloscopes that offer fast and reliable solutions for everyday test and measurement tasks such as embedded design development, power electronics analysis and general debugging. They are available with bandwidths from 200 MHz to 1 GHz. The scopes' highly accurate digital trigger system with virtually no trigger jitter delivers precise results. The single-core A/D converter with more than seven effective bits (ENOB) almost completely eliminates signal distortion. With a sampling rate of 5 GS/s and a maximum memory depth of 50 Msample per channel, the R&S RTE oscilloscopes can accurately record the long signal sequences required when analysing the data content of serial protocols such as I2C and CAN.

For further details: Ph: 011-42535400, sales.rsindia@rohde-schwarz.com, www.rohde-schwarz.co

COMPONENTS

Current sense resistors

In May 2014, **Vishay Intertechnology** introduced its surface-mount, four-terminal power metal strip Â current sensing resistors in the compact 2726 and 4026 case sizes to provide an operating temperature range of -65°Â C up to +275 Â°C. The Vishay Dale WSLT2726 and WSLT4026 combine their high-temperature performance

with 3 W power ratings and extremely low resistance values down to 0.002 Î. They incorporate a solid metal nickel-chrome alloy resistive element with low TCR (< 20 ppm/Â°C). This results in high-power resistors that offer an operating temperature range of -65 Â°C to +275 Â°C while maintaining the superior electrical characteristics of the power metal strip construction. The resistors are ideal for all types of current sensing, voltage division and pulse applications.

For further details: Ph: 080-25586277, business-asia@Vishay.com, www.vishay.com

Solar micro inverter development kit

In May 2014, **Texas Instruments** launched its C2000 solar micro inverter development kit. The kit implements a complete grid-tied solar micro inverter based around TI's C2000 Piccolo TMS320F28035 microcontroller (MCU). Solar micro

inverter systems place smaller or micro inverters at the output of each individual solar panel. This configuration offers benefits like the elimination of partial shading conditions, increased system efficiency, improved reliability and greater modularity. It supports panel voltages of 28 V to 45 V at input as well as universal power output, at up to 280 W for 220V AC and up to 140 W for 110V AC, making it suitable for the diverse requirements of solar markets, worldwide.

For further details: Ph: 080-25586277, business-asia@Vishay. com, www.vishay.com

Smart-meter ICs

In May 2014, **STMicroelectronics** unveiled its smart-meter ICs for the low-energy age in which billing is done by measuring accurately down to extremely low power levels. Meters featuring ST's new STPM32, STPM33 and STPM34 ICs will help utilities minimise revenue losses and

ensure consistent billing for even the most frugal customers. The STPM3 devices prevent losses by keeping the meter accuracy down to just a few milliamps, comparable to the current drawn by an LED television in standby. By performing power-quality calculations on-chip, including RMS voltage and current measurement, apparent-energy computation and under-voltage/over-voltage detection, the chips can offload the meter's host processor, thus simplifying software.

For further details: Ph: 080-66514000, www.st.com

Wire-to-board and wire-to-wire connectors

In May 2014, **FCI** launched Minitek Pwr 3.0 and Minitek Pwr 4.2 wire-to-board and wire-to-wire connectors. Minitek Pwr 3.0 is designed for power applications with current ratings of up to 5A per contact. Engineered for wire-to-wire and wire-to-board applications, its crimp and snap-in receptacles are used to terminate AWG 20 to 30 wires. Headers assemblies for wire-to-board interconnections include vertical and

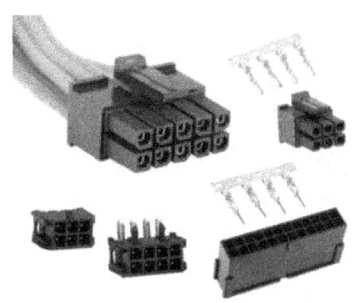

right-angle variants. Minitek Pwr 4.2 is manufactured for high-current and high-density applications, and supports up to 9A per contact. Both the products are designed for dual-row and 2 to 24 circuits, but feature greater versatility due to their diverse wire-to-wire, wire-to-board and wire-to-panel mount capabilities. Its crimp and snap-in receptacle is used to terminate AWG 16 to 28 wires.

For further details: Ph: 080-2559 7149/2555 0852, communications@fci.com, www.fci.com

LDO line-up

In May 2014, **ROHM** introduced a new 16-model line-up of LDOs optimised for MCU power supplies in automotive body and power train systems. The new power supplies in the BD4xxMx series, together with the BDxxC0A series, have been designed for application in car infotainment systems, bringing the total number of automotive-grade LDOs ROHM offers to 43. The BD4xxMx series utilises state-of-the-art power system 0.35µm BiC-DMOS processes and takes advantage of ROHM's re-

nowned analogue design technology to achieve less than half the no-load current consumption of standard products, contributing to significant energy savings. In addition, a novel circuit design enables support for ceramic capacitors, eliminating the need for electrolytic capacitors to prevent oscillation, reducing both the mounting area and costs.

For further details: niranjan@ rohm.com.sg, www.rohm.com

Power converter

In April 2014, **Mornsun** launched a higher power Din-Rail AC-DC power supply—the LI120, following the LI24 (24W) / LI72 (72W). It acts as the power supply for the industrial bus in the control cabinet, can provide all units with 24 V voltage, and can be used in industrial control / electric and other distributed power systems. The LI120 adopts a new circuit design and embeds power factor cor-

rection, which takes its efficiency up to 92 per cent, while pushing its standby power consumption down to 0.75 W. With 35mm installation width, it saves considerable space for users. In addition, this Din-Rail mounting product protects a circuit from the damages caused by over voltage, overload or continuous short circuits.

For further details: www. mornsurn-power.com

POWER ELECTRONICS

Tubular batteries

In May 2014, **Base Batteries** launched the Base Tuff BT 500 tall tubular batteries for home UPS systems. These tubular batteries are supposedly the most energy efficient backup units

designed to protect homes from the inconvenience caused by power outages, minimise energy consumption and help consumers cut down on electricity bills. The tall container allows for a greater amount of electrolyte, which reduces maintenance and increases battery life even in tough power conditions. The BT series is environment-friendly and free from acid fumes. It has a water level indicator and a special tubular positive-plate design for long life in the deep discharge cycle.

For further details: Ph: 080-41635909, helpline@basecorporation.com, www.basebatteries.com

Home UPS system

In April 2014, **Luminous Power Technologies** introduced Zelio, a pure sine wave, state-of-the-art home UPS system, designed to work with the most sensitive home appliances. Zelio has a unique feature that displays the

power back-up in hours and minutes, which allows consumers to plan their day and manage their appliances during a power outage. This product can be used to provide power backup for all domestic appliances and is equipped with various in-built protection features. This system with next generation architecture by Luminous comprises a digital signal processing core that yields a high performance.

For further details: mktg@lumnousindia.com, www.luminousindia.com

CHEMICALS & CONSUMABLES

High-reliability solder paste

In April 2014, **Indium Corporation** introduced the SACm, a highly reliable solder paste that increases the drop-shock performance in portable electronics by 800 per cent, without compromising on thermal cycling.

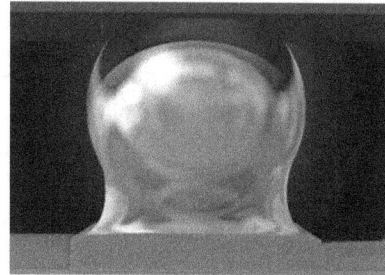

SACm is doped with manganese and contains less silver than other lead-free solder pastes. The manganese provides increased strength and the reduced silver content provides a more stable cost structure, especially beneficial for cost-sensitive applications. Indium8.9HF halogen-free, lead-free solder paste is one of the standard flux vehicles for SACm that provides excellent soldering performance under high temperatures and long reflow processes. It offers unprecedented stencil print transfer efficiency and works in a wide range of processes.

For further details: Ph: + 65-6268-8678, asiapac@indium.com/sacm@indium.com, www.indium.com/SACm

SECURITY & SURVEILLANCE

Dome panoramic camera

In May 2014, **Axis Communications** launched the compact AXIS M3027-PVE, a 5-megapixel fixed dome camera that offers a 360° or 180° overview of wide areas. The IK10 vandal resistant, day/night network camera

is designed for indoor as well as outdoor use. It can be mounted on ceilings, soffits or under porches for a 360° overview and on walls for a 180° view. The camera provides de-warped views such as panorama, double panorama and quad views, as well as view modes where users can digitally pan, tilt and zoom in on areas of interest. Multiple streams in H.264 and Motion JPEG can be sent simultaneously.

For further details: Ph: 080-41571222, www.axis.com

MISCELLANEOUS

Modular surface mounter

In July 2014, **Yamaha Motor Co Ltd** launched its new high-efficiency modular surface mounter, the Z:LEX YSM20. This new surface mounter has the versatility to cater to a variety of different production processes, whilst supposedly achieving the world's fastest mounting speed in its class (under optimum conditions) at 90,000 CPH. The Z:LEX YSM20 combines features found in four of Yamaha's current models—the 2-beam, 2-head compact super-high-speed YS24 modular mounter; the compact, high-speed flexible YS24X modular mounter; the single-head, single-beam high-speed general-purpose YS100 modular mounter; and the multi-functional wide-range YS88 modular mounter. The new model also allows production changeovers to be carried out with ease.

For further details: India distributor: Transtechnology India Pvt Ltd, Amit Madan, country manager–India, Ph: 09810449898

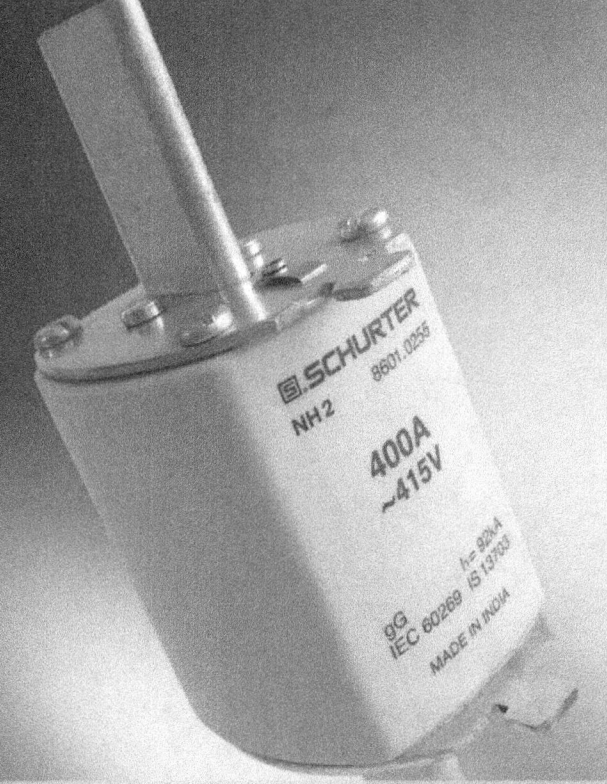
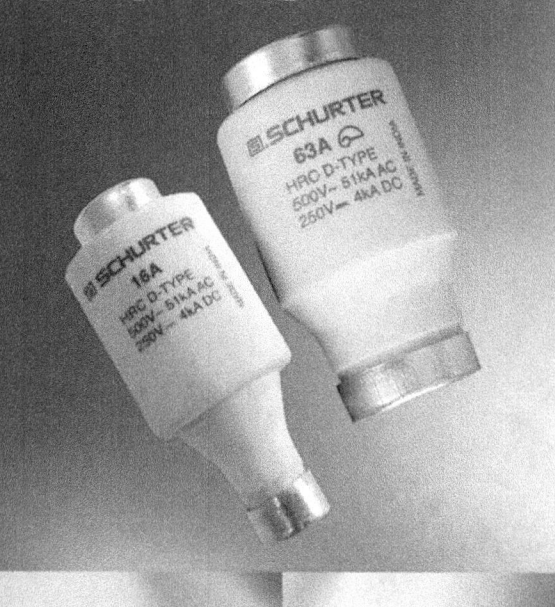
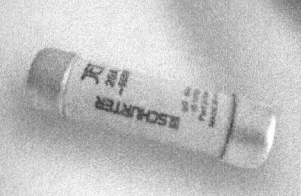
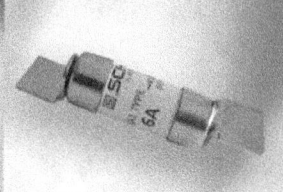

High-mix thinking
for a higher
volume world

Volume to NPI and back. Without missing a beat.
The new MY200 performance series.

Now you can leapfrog between varied batch sizes in the blink of an eye. Having long developed flexible solutions for the demanding aerospace industry, we understand the needs for agile, high-quality placement of vital components. Success is not just a matter of machine speed, but how many boards you can mount at the end of the day.

This high-mix thinking translates well to todays automotive and other higher volume industries where batch sizes are decreasing. Why? Because car electronics are becoming as diverse and advanced as the models themselves. So why not put the industrys most agile solution to the test in a demo of our new, improved MY200 series? See how it can improve your production quality, with zero mix-up of parts, streamlined material handling, and traceability down to individual placements. Give us a call to find out how our high-mix thinking could benefit you.

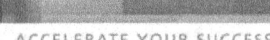

MYDATA®

ACCELERATE YOUR SUCCESS

www.mydata.com

www.electronicsb2b.cor

The new MY200 performance series.

higher throughput, higher accuracy and improved line utilization

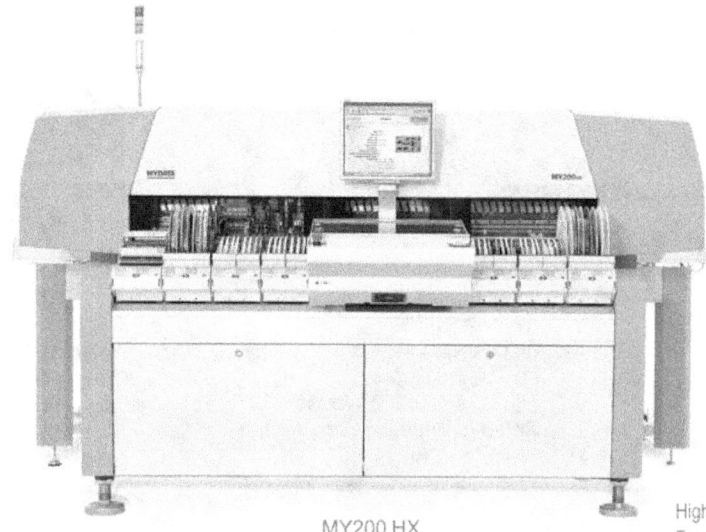

MY200 HX

High-speed all-in-one mounter.
Featuring up to 160 feeder positions.
Top speed of 40,000 components per hour.

MY200 DX

High-speed chip and low-profile IC mounter.
Featuring up to 160 feeder positions
Top speed of 50,000 components per hour.

MY200 SX

Flexible all-in-one mounter.
Featuring up to 176 feeder positions.
Top speed of 24,000 components per hour

MY200 LX

All-in-one mounter.
Featuring up to 176 feeder positions.
Top speed of 16,000 components per hour.

for details CONTACT:

ACCUREX

Accurex Solutions (P) Ltd.
782, 3rd Main, 1st Cross, 5th Phase, BEML Layout,
Rajarajeshwari Nagar, Bangalore - 560 098. INDIA.
Tel : +91-80-2861 1871
TeleFax : +91-80-2861 2021
Email : info@accurexsolutions.com

DELHI	: delhi@accurexsolutions.com
JAIPUR	: jaipur@accurexsolutions.com
MUMBAI	: mumbai@accurexsolutions.com
HYDERABAD	: hyd@accurexsolutions.com?
CHENNAI	: chennai@accurexsolutions.com
MYSORE	: info@accurexsolutions.com

www.accurexsolutions.com

 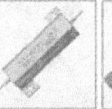 

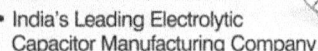

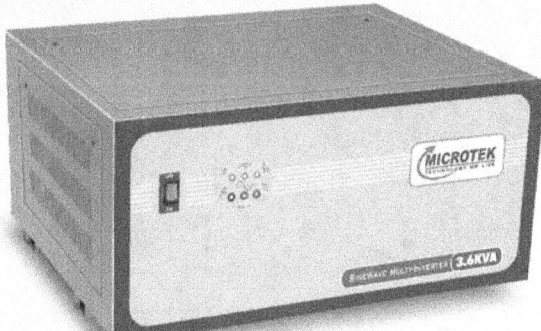

SILICON POWER ELECTRONICS
D-2,29/3,MIDC,Chinchwad,Pune-411019, INDIA   M:9011013275   spe@vsnl.com

Manufacturer of Power Semiconductor Devices :

Rectifier Diode

Thyristor

Single / Three Phase
Bridge Rectifier

Single Phase
SCR based rectifier

Thyristor Modules

Diode Modules

Power Zener Diode

Solid State Relay

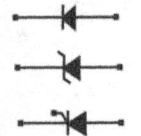

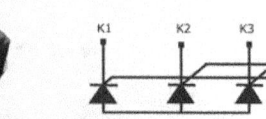

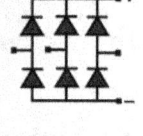
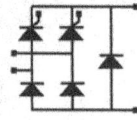

YOUR PARTNER FOR HIGH QUALITY PRINTED CIRCUIT BOARDS

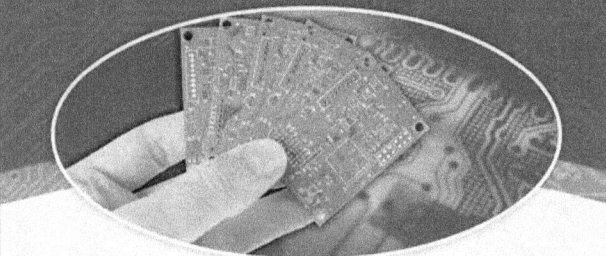

CUSTOMER SUCCESS IS OUR SUCCESS

Quick and easy access through Internet and E-mail. With our strong front end engineering we are able to help many customers to grow faster and better.

COMMITMENT: On time delivery of Quality PCB's at significantly lower than European prices, competitive with World-wide prices

CERTIFICATION: ISO 9001:2008, C-DOT, LCSO, UL (File E 135990)

QUALITY STANDARDS: IPC Class 2, 3/perfag/JSS/LCSO/Customer Specification

Turnkey solutions to core sectors like - Automotive/Aero/Defense, Computing /Storage, Green Technologies, Industrial Security, Medical, Networking/ RF/ Wireless, Metering/Power, Lighting etc… in India and abroad.

Bharat Electronics Limited, Pune awarded SCL with "ZERO DEFECT SUPPLIER AWARD" for the year 2011-12 in their Supplier Meet – 2012.

SULAKSHANA CIRCUITS LTD
Plot No. 36&37, Anrich Industrial Estatel.D.A. Bollaram, Hyderabad - 500 032, Andhra Pradesh, India. Tel: +91-040-32420340, Email: info@sclcircuits.com

---

SAVE POWER SAVE NATION

PACIFIC's Power Saver
Model: PPS-216

Automatic Alternate Operation of AC and Fan

Technical specifications:
• The power saver is connected in series to AC and ceiling fan.
• Power consumption of Power saver is 1mV maximum.
• Output control is done through a contactor of rating 32amp.
• Time control is done through Microcontroller.

PACIFIC Pacific Electronics Pvt Ltd.
Plot no-22, Ground Floor, Ravi-Co-op. Society, Opp. road to SBI Trimulgherry Branch, Trimulgherry, Secunderabad- 500015
Tel No.: 040-27797131, 27791159, Mob: 9912341728
Email: audiogen@usa.net

Dealers and Distributors inquiry solicited

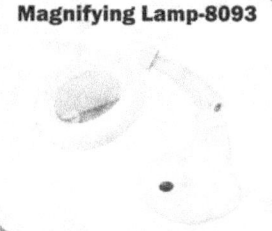

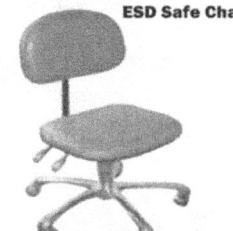

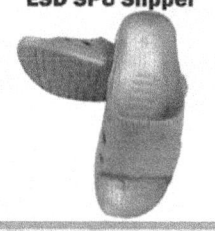
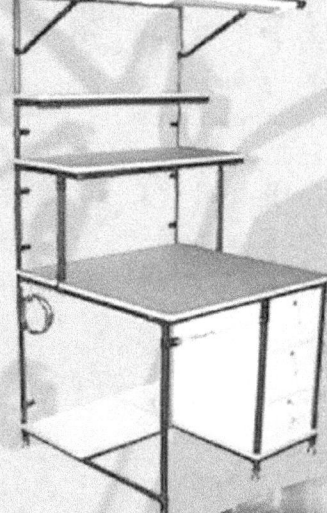

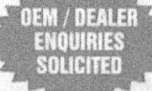

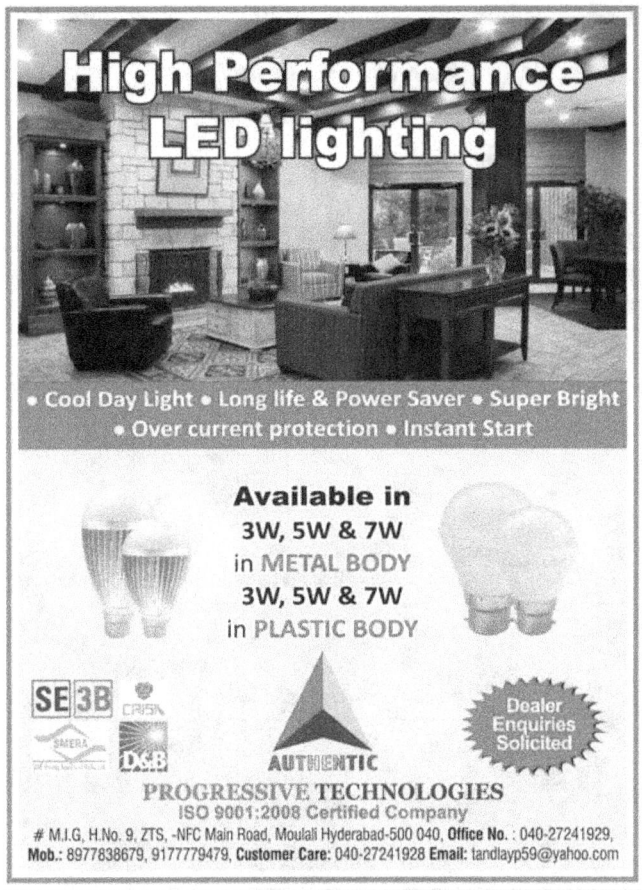

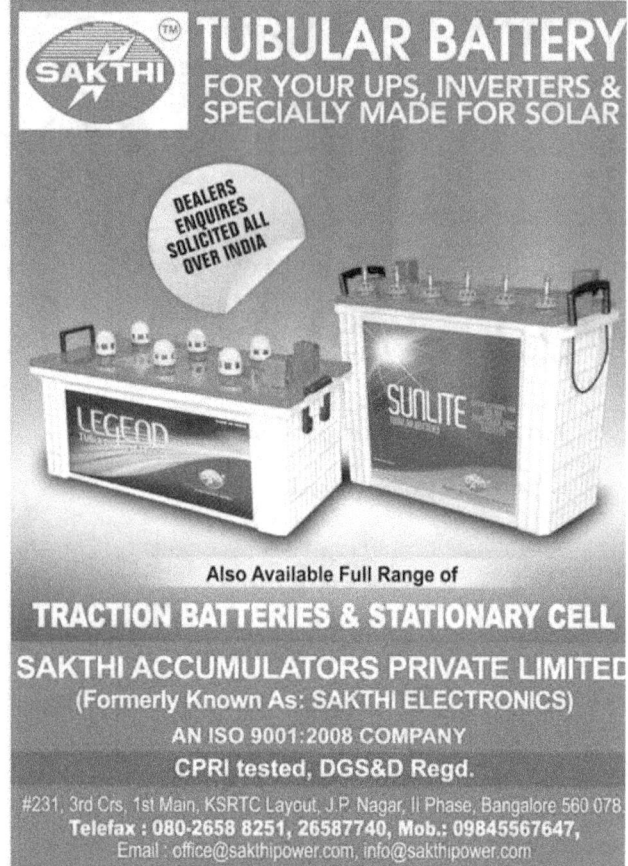
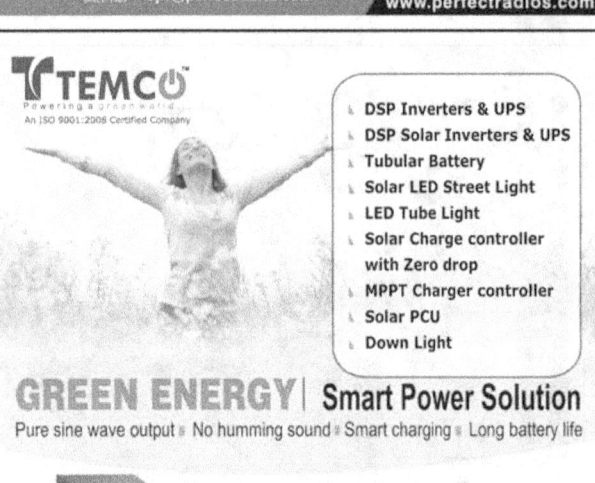

WORLD's BEST CUTTERS FROM USA

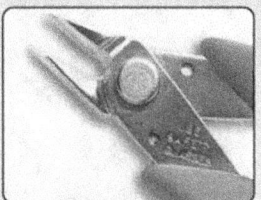

LX- Micro-Shear Flush Cutter

Our patented Micro-Shear cutting action combined with precision ground cutting edges, extra tough 100" thick high carbon steel blades, an ultra slim profile for access in high density areas. Sized for smaller hands and maximum maneuverability. Flush cuts soft wire up to 16 AWG (1.29 mm).

Note: ESD models are also available

LX-F Micro-Shear Flush Cutter

Our LX with a factory installed lead retainer. Design provides a low profile and non-slip mounting on the shear. Helps prevent flying leads and Component shorting from stray leads.

Note: ESD models are also available

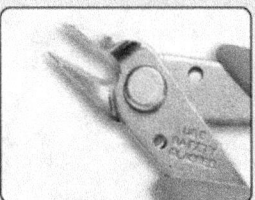

LX -T Micro-Shear Flush Cutter

A special extra tapered cutter head with a ultra-slim tip profile combined with the same Micro-Shear cutting action, precision grinding, and tough high carbon steel blades of the standard LX. Ultra-Sharp, precision tip for access in highly restricted areas. Flush cuts soft wire upto 20 AWG (0.8mm)

Note: ESD models are also available

2175 Maxi-Shear Flush Cutter

An extremely durable and versatile wire cutter featuring our patented Micro-Shear® cutting action. Tough enough for harnesses and cables with the precision to cut material less than 1 mil, thick or work in high density areas. Ergonomically shaped, non-slip Xuro-Rubber™ grips and a glare-eliminating black finish for operator comfort. Non-protruding, lifetime warranted flat spring provides excellent "feel" without excessive spring tension. Flush cuts soft wire up to 12 AWG (2.05mm).

Note: ESD models are also available

170-IIF Micro-Shear Flush Cutter

The 170-IIF features a low profile, non-slip, factory-installed lead retainer to help prevent flying lead and component shorting from stray leads.

Note: ESD models are also available.

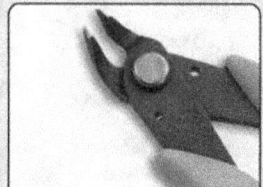

280-II Angled Micro-Shear Flush Cutter

The 280-II allows easier cutting in difficult areas. A 55º angled head provides excellent sight lines when used in either the vertical or horizontal position. A slim profile increases accessibility to hard to reach surfaces. Light weight, glare eliminating black finish, and non-slip Xuro-Rubber™ grips ensure operator comfort. Flush cuts soft wire up to 20 AWG (0.8mm).

Note: ESD models are also available

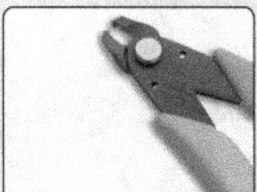

670 Cut & Crimp Tool

The 670 Cut and Crimp tool features a slim profile (0.085") that allows accessibility in high density areas and provides excellent sight lines for precise positioning. Swaged leads are firmly attached to PCB, while freshley exposed lead material enhances solderability. High Carbon Steel guarantees durability. Cuts & Crimps soft wire upto 20 AWG (0.8mm)

Best results are achieved when the diameter of the hole on the PCB is not larger than 2.5 times the diameter of the wire being cut/crimped.

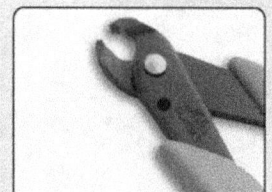

670HD Heavy Duty Cut & Crimp Tool

The 670HD offers the same great features as our 670 Cut and Crimp tool in heavy-duty configuration. A larger head (0.115") simplifies positioning and heavy-duty construction provides extended tool life and larger wire capacity. Light weight, glare - eliminating black finish and ergonomically – shaped, non-slip Xuro-Rubber™ grips ensure operator comfort. Cuts and crimps soft wire up to 16 AWG (1.27 mm).

Best results are achieved when the diameter of the hole on the PCB is not larger than 2.5 times the diameter of the wire being cut/crimped.

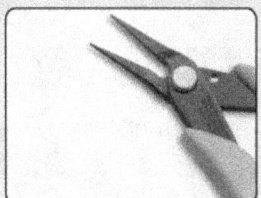

450 Tweezer Nose Plier

The 450 is an ultra-precise needlenose plier capable of grasping and holding wire less than 1 mil thick with the strength for wire forming. Leads "popping" free, tip mis-alignment and blade crossover will be a memory once you switch to out Tweezer NoseTM pliers. Radiused edges protect lead wires. Light weight and our patented, non-protruding, Light TouchTM return spring help ensure operator comfort.

Note: ESD models are also available

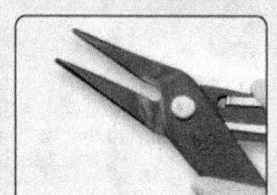

485 Xuro-Grip Long Nose Plier

A true "electronics production long nose plier" is what we had in mind when we designed the 485. Engineers and production managers told us traditional drop-forged designs were too bulky and lacked the sensitive "feel" required for precision electronic assembly. Our 485 performs where the competition failed. 485 features a thin profile for access in high density areas, radiused outside edges, ergonomically-shaped, nonslip Xuro-RubberTM grips, a glare-eliminating black finish and light weight for operator comfort.

Note: ESD models are also available

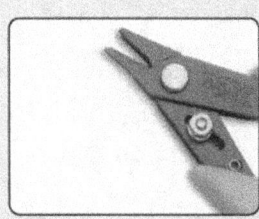

573 Xuro-Former Lead Forming Tool

The patented, original strain-relief forming hand tool. Radius and depth of bend are controlled by a simple slide adjuster, which locks firmly in place to privent unintended setting changes. this high cabon steel tool features light weight, glare-eliminating black finish and ergonomically shaped, non-slip xuron rubber grips. Forms lead wires upto 0.030" (0.8mm)

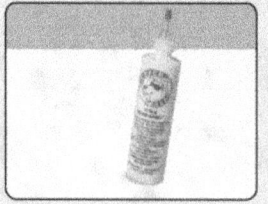

Rover Mask 775-8

This convenient 8 ounce (227g) applicator bottle features a tapered spout for easy application. Inner safety seal keeps mask fresh and prevents spillage in transit.

Note:Rover mask is also available in a one gallon (8.25lbs / 2.75kg) The most economical way to purchase in 55 gallon drum (Sold by weight)

STATIC SYSTEMS
TOTAL SOLUTION FOR ELECTRONIC INDUSTRIES

Head Office:
"Sri Krupa", No. 925/3, I Floor, 3rd Main, 3rdCross, 2ndStage, 'D'Block, Rajajinagar, Bangalore-560 010.
Ph: 91-80-4128 5756 / 91-80-23133597, Telefax: 91-80-23125475,
E-mail: support@staticsystem.com, chandru@staticsystem.com, Website: www.staticsystem.com

Delhi Office:
Flat No 315, 2nd Floor, PKT-1, Sector 14,
Dwarka, New Delhi - 110075
Ph: +91 9811328157 Email: sales@staticsystem.com

www.staticsystem.com

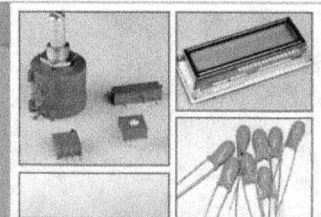

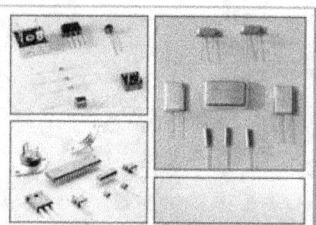

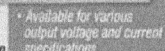

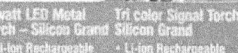
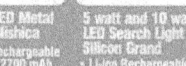
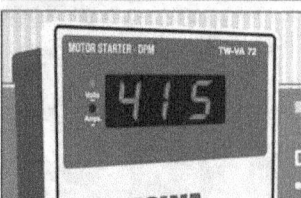

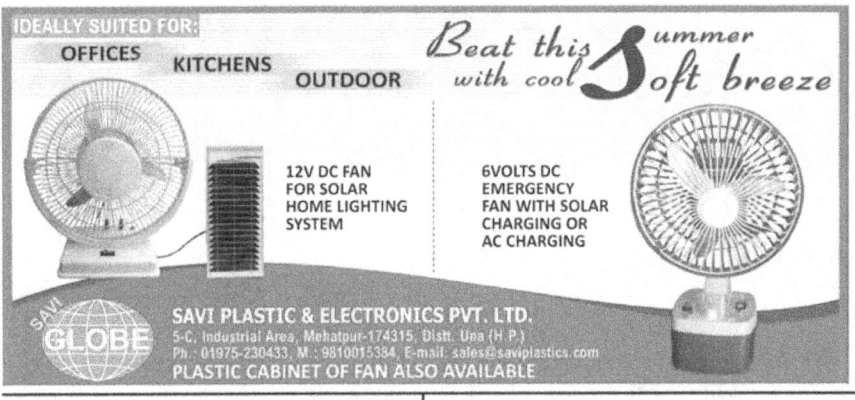

Omron Automated Optical Inspection

Maximize your SMT Line throughput with the use of quick setup and highly accurate inspection capability.

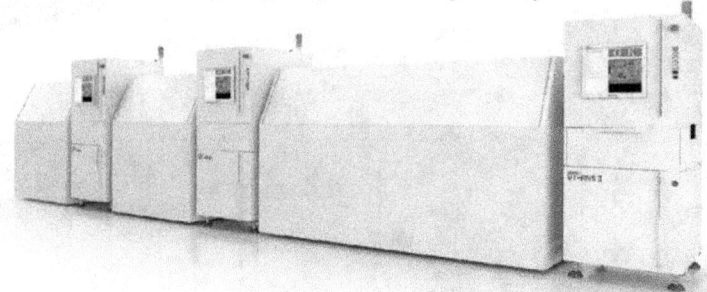

Introducing 2nd Generation RNS Series

VT-RNS2 ptH
(Bench Top)

VT-RNS2
(In-Line)

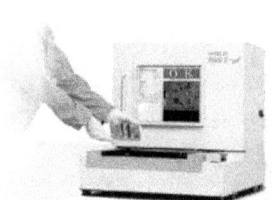

- High Resolution Inspection using 3CCD Camera True Color image proceesing – (RGB)
- Inclusive New S/W's – Fast & Easy Programming – User friendly
- Line Status Reporting, Full traceability
- Lead-Free enabled.

The NEW Generation3D Inspection System

VT– S500 Inline Pre & Post Reflow PCB Inspection (3D)

VT–S720 Inline Post Reflow PCB Inspection (3D)

- Realizing Vertical High Speed Startup with Auto Programming & Stable Inspection (IPC Compliant Program setting).
- Flexible access to the tools by Web Application.
- Real Time Process Monitoring & Full proof Traceability with Quality Control
- Dual Lane Option (No increase in Foot print)
- User friendly software – Touch Screen Interface

- Realizing Vertical High Speed Startup with Auto Programming & Stable Inspection (IPC Compliant Program setting).
- Flexible access to the tools by Web Application.
- Multi Angled, Oblique Cameras
- Real Time Process Monitoring & Full proof Traceability
- User friendly software – Touch Screen Interface

3D SPI – Inline paste Inspection

VT-X700 X-Ray Inspection System

- Unique Concept of 3D Measurement with Touch Screen Pannel with User–friendly software / Operations.
- Fast & Easy Programming in 3 min.
- Automatic Warpage Compensation Automatic Resolution setting as per program need, makes faster & better Inspection.
- Most Advanced SPC Software with detailed report function.
- Closed Loop Function up with Feedback & Feed forward link with Printer & Mounter.

- High-Precision, X-Ray, Angled 3D CT Imaging.
- Safe and Secure Closed X-Ray Tube
- User-Friendly Software and Interface, Fast and Accurate Automated Inspection.
- Automatic Component Model Creation for Quick and Easy Programming
- Defect Checking Terminal and 3D Image Reviewing Software.
- Data Analysis and Quality Control Software Tool.
- Pulse–Shot X-Ray Method for Extended X-Ray Source Life.

Omron Q-upNavi system – A True Process Improvement Tool

- The information provided by Q-upNavi results in number of outcomes.
- Navigates Engineers and Operators in locating all Process Defects.
- Quick Result Feedback of Manufacturing Process Changes.
- Quality Analysis of Process Defects, supported by High Quality Colour Images.
- Stabilizing Process providing Achievable Process Target.
- Reduction in AOI Optimizations and Verification & Reduction in Material Waste.

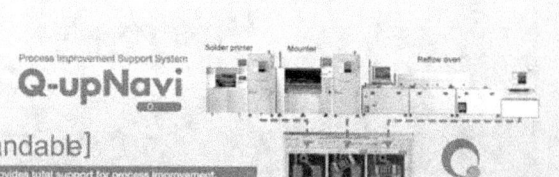

Q-upNavi

[Expandable]

Q-upNavi provides total support for process improvement, root cause defect analysis and countermeasure implementation

Q-upNavi is quality control software that analyzes inspection results and provides feedback to the production line. This software enables operators to implement corrective procedures that will prevent future defects from occuring regardless of their level of experience or expertise.

OMRON AUTOMATION PVT. LTD. 5th Floor, 501 – 504, The Qube, M.V. Road Marol, Andheri (E), Mumbai – 400059 T +91 22 71288400 ext 13
www.omron-ap.co.in Inquiry : samitag@ap.omron.com
Contact : Hemant Gangurde – Mobile: 9167930563 Tel : (91) 22 71288473

Authorised Distributor: Leaptech Corporation
812 Cosmos, Sector 11, CBD Belapur, New Mumbai 400614, India Tel : (91) 22 2756 2822 URL : http://www.leaptech.in Inquiry : leaptech@vsnl.net

R N I No. DELENG/2007/20027, Mailed on 14-15th of Advance Month   Delhi Postal Regd. No. DL(S)-01/3316/2014-16
Published on 13th of the Advance Month   Licenced to Post without Pre-Payment Licence No. U(S)-32/2014-16

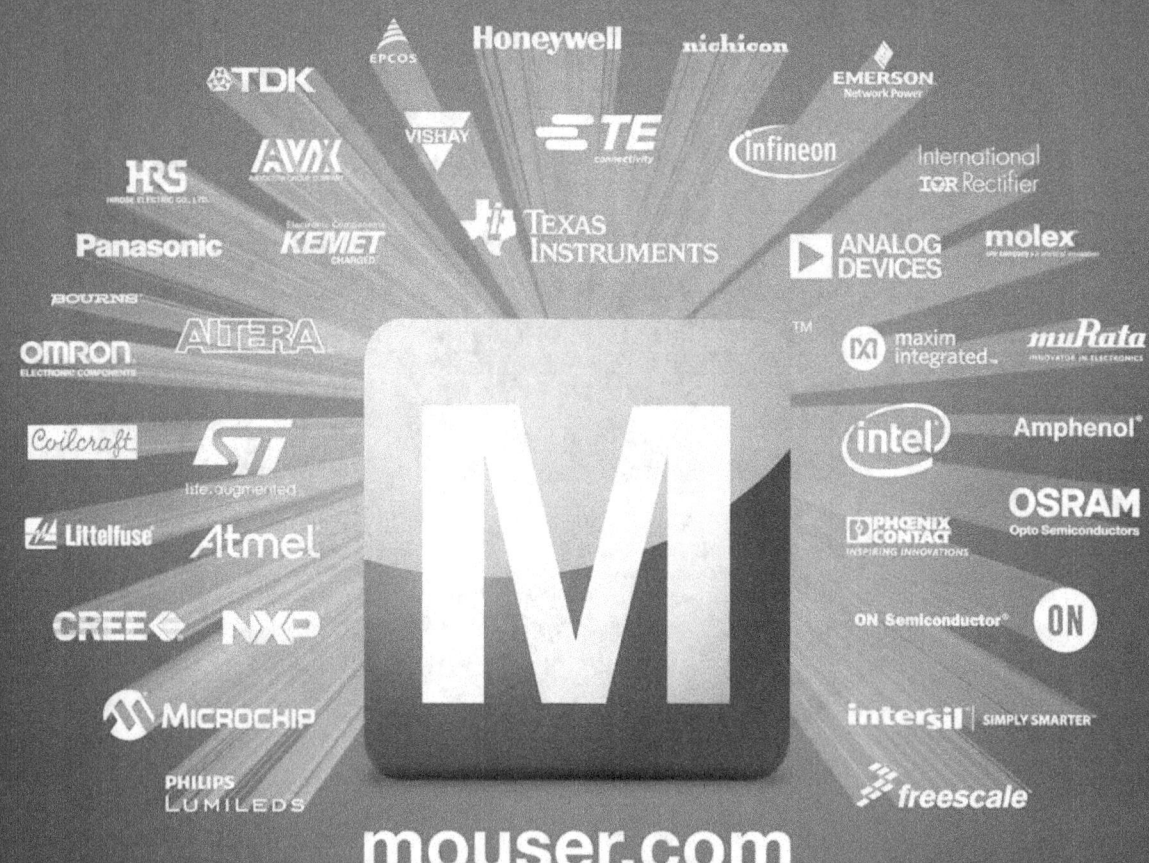